Edgar Fahs Smith, John Marshall

Chemical Analysis Of The Urine

Edgar Fahs Smith, John Marshall

Chemical Analysis Of The Urine

ISBN/EAN: 9783741134456

Manufactured in Europe, USA, Canada, Australia, Japa

Cover: Foto ©Angelika Wolter / pixelio.de

Manufactured and distributed by brebook publishing software
(www.brebook.com)

Edgar Fahs Smith, John Marshall

Chemical Analysis Of The Urine

CHEMICAL ANALYSIS

OF

THE URINE,

BASED IN PART ON

(CASSELMANN'S ANALYSE DES HARNS,)

BY

EDGAR F. SMITH, Ph.D.,
Asa Packer Professor of Chemistry
in Muhlenberg College.

JOHN MARSHALL, M.D.,
Demonstrator of Chemistry, Medical Department, University of Penna.

WITH

ILLUSTRATIONS.

-- --

PHILADELPHIA:

PRESLEY BLAKISTON,

1012 WALNUT STREET.

1881.

PREFACE.

Intimate association with students as instructors in medical chemistry has revealed to the authors the fact that none of the existing works on urinary analysis deal sufficiently with the chemical side of the subject. Cognizant of this, and believing that the requirements of the present curriculum demand a more thorough knowledge of details than is usually presented, we have endeavored to collect in the following pages all matter bearing on the chemical analysis of urine which experience has demonstrated to be practical and thoroughly reliable. Selecting as a partial basis for our work the admirable little publication of Casselmann—*Analyse des Harns*—we have added numerous methods of analysis and suggestions to enable the student at work in the laboratory, or privately, to perform understandingly the solution of the many problems met with in the analysis of urine.

As volumetric methods of analysis are readily applied in estimating the urine constituents, the preparation of standard solutions and the accompanying calculations have received due attention. Following immediately upon the close of the chemical portion of the work will be found a section upon the microscopic examination of urinary sediments, interesting alike to the student and practitioner of medicine.

The plates illustrating the microscopic character of various urine constituents are borrowed from Casselmann, while the apparatus of Hüffner, for the estimation of urea, is here introduced from the *Journal für prakt. Chemie.* Its simplicity and accuracy recommend its general adoption. For the apparatus pictured in the frontispiece, we are under obligations to Prof. Wormley, to whom, as well as to Prof. Samuel P. Sadtler, we would here express our sincere thanks for the many kindnesses shown us during the progress of our labors.   S. AND M.

CONTENTS.

I.

vii

IV.

ABNORMAL CONSTITUENTS OF URINE; THEIR OCCURRENCE AND DETECTION.

V.

URINARY DEPOSITS (SEDIMENTS).

VI.

PRACTICAL HINTS TO A COURSE FOR THE QUALITATIVE AND QUANTITATIVE EXAMINATION OF URINE.

VII.

URINARY CONCRETIONS.

CHEMICAL ANALYSIS

OF

THE URINE.

I

THE URINE.

The urine is that peculiar fluid eliminated by the kidneys, in which we find the elements that have become useless to the animal economy in the form of soluble nitrogenous bodies and salts. We can distinguish two varieties of urine among the mammals, depending entirely upon their nourishment, viz :—

(*a*) Urine of Herbivorous animals.
(*b*) Urine of Carnivorous animals.

The first is characterized by its constant *cloudy appearance*, its *alkaline reaction* and the remarkably large quantity of phosphates of the alkalies, and alkaline earths present in it. Uric acid is entirely absent, while hippuric acid is abundant in it. The urine of carnivorous animals in a fresh condition is *clear*, light yellow in color, with an agreeable odor, bitter taste (due to urea and indican. K. B. Hofmann), and *acid* reaction. It is rich in urea, but almost perfectly free from uric acid, which at the most occurs in traces.

2. Normal human urine resembles the second variety

(urine of Carnivoræ). Freshly eliminated it is clear, of an amber yellow color, with a decided acid reaction (due, according to Liebig, to acid phosphates; according to Lehmann, to free hippuric and lactic acids), and a bitter, saline taste and peculiar odor (arising from phenol. Städeler). Its specific gravity varies from 1.005 to 1.025, depending upon nourishment, sex, age, etc.

3. Its normal and constant constituents are water, urea, uric acid, hippuric acid, creatin, creatinin, xanthin, coloring matters, indican, mucus from the bladder, chlorides, phosphates and sulphates of potassium, sodium, ammonium, calcium and magnesium, and now and then traces of iron, nitrates and silica. The gases present are nitrogen and carbon dioxide. The quantities of these substances are variable, and frequently they occur in such minute traces as to render their estimation very difficult.

4. The behavior of urine with chemical reagents may be briefly outlined as follows: On boiling, normal urine should remain clear, and generate, when mixed with concentrated acids, a peculiar, nauseous odor, and at the same time become more or less dark in color. Immediate cloudiness does not ensue, but in course of time crystals of uric acid separate.

The *alkalies* precipitate the phosphates of the alkaline earths (calcium and magnesium phosphates).

Barium Chloride in urine acidified with hydrochloric acid precipitates sulphuric acid as barium sulphate.

Silver Nitrate in urine acidulated with nitric acid, throws down silver chloride. (If the acidulation be omitted silver phosphate will also be precipitated.)

Ferric Chloride precipitates the phosphoric acid from urine previously acidified with acetic acid.

Lead Acetate precipitates the chlorides, sulphates and phosphates as lead salts.

Oxalic Acid or *Ammonium Oxalate* precipitates calcium as oxalate.

Mercuric Nitrate produces in urine, after the removal of sulphuric and phosphoric acids, at first a cloudiness, which disappears, caused by the following reaction:—

$$Hg(NO_3)_2 + 2Na\,Cl = Hg\,Cl_2 + 2Na\,NO_3.$$

When this change—the conversion of sodium chloride into nitrate, and mercuric nitrate into chloride—is completed the further addition of mercuric nitrate will induce the separation of a white insoluble compound of mercuric oxide and urea.

Alcohol produces a cloudiness, which disappears upon dilution with water.

5. After protracted standing normal urine undergoes a change; fermentation begins, and this is either—

(*a*) Acid fermentation, and afterward
(*b*) Alkaline fermentation.

According to Scherer, the mucus from the bladder contained in the urine decomposes, forming a fungus very similar to the ferment (Mycodermæ Cerevisiæ), and then it decomposes the coloring matter that may be present. Usually the color of the urine grows paler in consequence, and yields a more acid reaction, due to the formation of lactic and acetic acids, and in addition red-colored sediments (mixtures of uric acid, urates and mucus) deposit out. From this we observe that the acid fermentation of urine stands in close relation to the formation of urinary deposits and the production of calculi.

6. Gradually the urine, dependent on the temperature, the cleanliness of the vessels, etc., passes from the acid into

the alkaline fermentation.   Indeed, it is not absolutely necessary that the acid fermentation should have preceded this; as, under certain circumstances not yet explained, the urine enters into the alkaline fermentation in the bladder. Here it is induced by the mucous coating of the bladder (according to Tiegheim and Schönbein by distinct, peculiar fungi (Torulaceæ), and it is for this reason that we observe, in affections of the mucous membranes of the bladder, that the urine that has been recently passed possesses an alkaline reaction.

In the alkaline fermentation of the urine the urea decomposes into · acid ammonium carbonate and free ammonia—

$$CO(NH_2)_2 + 2H_2O = NH_4HCO_3 + NH_3.$$

We notice, in consequence, a strong ammoniacal odor, and also that upon the addition of acids to the liquid, strong effervescence ensues.   The ammonia liberated unites with the magnesium phosphate and produces the so-called *triple phosphates,* which separate as a microscopic crystalline precipitate.   Their form (coffin-lid shape) is characteristic. In most cases there is a simultaneous formation of a thin coating upon the surface of the urine, and besides, with the assistance of a microscope, fungus threads, with and without spores, infusoria (vibrionæ and monadæ), and ammonium urate are observed.   Mixed with alkalies there follows an abundant generation of NH_3.

Abnormal Constituents of Urine.

7. Albumen, glucose, alkapton, inosite, lactic acid and lactates, fats and volatile fatty acids, benzoic acid (usually converted into hippuric acid), succinic acid, biliary coloring matters, biliary salts, allantoin, leucin, tyrosin, cystin,

taurin, mucin, hæmatin, fibrin, pus, spermatozoids, ammonium carbonate, triple phosphate and hydrogen sulphide.

8. Substances that have been detected in urinary deposits are: uric acid, urates, calcium oxalate and phosphate, ammonium magnesium phosphate, ammonium carbonate, cystin, tyrosin, xanthin; and of organized substances: mucus and epithelia, pus, blood and spermatozoids, fungi and infusoria, fibrin, coagula, sarcinia ventriculi, Goodsir.

Substances that have received the designation " accidental constituents " are those which, by food or medicine, etc., have been introduced into the system and eliminated by the urine, partially changed or chemically altered in their form. The following have been detected in the urine, not altered by their passage through the system :—

(1) The majority of the salts of the heavy metals, when administered in rather large quantities. To this class belong the preparations of antimony, arsenic, mercury, zinc, gold, silver, lead, bismuth, etc.

(2) The alkaline carbonates, potassium iodide, ammoniacal salts.

(3) The free organic acids.

(4) The greater portion of the alkaloids.

(5) The greater portion of the dye and smelling substances.

The following have been found partially or entirely altered in their chemical nature: benzoic acid, quinic acid, cinnamic acid, and oil of bitter almonds as hippuric acid (therefore the occurrence of the latter with Herbivoræ):—

Tannic acid as gallic acid.

Alkaline salts of vegetable acids as alkaline carbonates.

Potassium sulphide as potassium sulphate.

Free iodine as potassium iodide.

9. Apparatus necessary in the examination of urine: Urinometer, a small alcohol lamp or Bunsen gas lamp, a water bath, wash bottle, twelve test tubes with stand, funnels, beaker glasses, porcelain evaporating dishes, watch glasses, glass rods, two to four pipettes à 5, 10, 20, and 50 c.c., a graduated cylinder with foot, filter paper, a polarization apparatus for the estimation of sugar, and Vogel's color scale.

10. The reagents that meet with most frequent use are, red and blue litmus paper, paper saturated with lead acetate, turmeric paper, paper saturated with ammonium molybdate, acetic, chromic, hydrochloric, nitric, fuming nitric, oxalic and sulphuric acids, ether, absolute and diluted alcohol, distilled water, fused silver nitrate and a solution of same, barium, calcium and ferric chlorides, Fehling's copper solution, fuchsin solution, mercuric nitrate, potassium or sodium hydrates, sodium acetate, carbonate, nitrate and phosphate, and zinc chloride.

II.

PHYSICAL PROPERTIES AND REACTIONS OF URINE.

11. The physical properties of urine which are of interest in diagnosis are the color, odor and specific gravity. In pathological conditions the normal amber yellow color of the urine is converted in some cases into a pale whitish yellow, and again to a red or brown black. Hence we distinguish as follows:—

(a) Pale urine—colorless to straw yellow.

(b) Normal color—gold yellow to amber yellow.

(c) High colored urine—reddish yellow to red.

(d) Dark urine—brown, dark beer color to black.

(*e*) Green urine.

(*f*) Dirty blue urine.

These different colorations lead us to the following conclusions:—

(*a*) The pale urine of patients would suffice to inform us that the affected individual was not suffering from any violent, acute, febrile disease. Yet its occurrence may be observed in convalescents, and many of those who have suffered from some chronic affection (anæmia, chlorosis, diabetes). Indeed, if long continued we can determine a certain degree of anæmia.

There are but minute quantities of coloring substances and urea in pale urine, and it is generally the case that the solid constituents are not abundant except in diabetes mellitus, and in healthy persons who drink much water or beer (urina potus).

(*b*) The normal colored urine justifies the conclusion that no sickness is present, in which either the pale urine or (*c*) occur.

(*c*) The highly colored urine, by its color and high specific gravity, proves conclusively that it is concentrated, rich in solid constituents, in urea, etc. The reaction is always acid. Persons in good health may, after the ingestion of rich food, eliminate a normal yet highly colored urine, but with sick persons the occurrence is of great importance to the physician, inasmuch as urine of this class accompanies all febrile diseases; in the case of hectic fever it forms a more positive guide than the pulse or temperature.

(*d*) Dark urine generally points to an abnormal pigment, which is present as an admixture in the urine, *e. g.*, biliary coloring matters, coloring matter of the blood, and

also uroxanthin. Not unfrequently the coloration is accidental, arising from medicaments like rhubarb, senna, carbolic acid and others.

(e) Green urine of a dirty hue arises from biliverdin,
in icterus, and brown icteric urine has the same import.

(f) Dirty bluish urine generally has a dark blue coating, and shows a blue deposit formed by the production
of indigo. The reaction is alkaline. This type of urine
is met with in cholera and typhus.

12. The odor of human urine has not yet been referred
positively to distinct chemical substances. It is merely
suspected that it is influenced or dependent upon extremely
minute quantities of phenylic, taurylic, damaluric and da-
molic acids (Städeler.) For the practitioner the odor of
the urine is of but minor significance, as it often varies in
consequence of the ingestion of foods, medicines, e. g., aspa-
ragus, oil of turpentine (violet odor), saffron, cubebs, and
similar substances.

In alkaline fermentation a disagreeable ammoniacal odor
is present. Heller observed, in cases of severe typhus and
spinal troubles, a peculiar musty odor, which indicated the
formation of fungi (possibly, the cause of the contagious
diseases).

13. The changes in specific gravity are worthy of consideration. The specific gravity of normal urine is greatly
influenced by the urea and sodium chloride, and can, according to J. Trapp, be employed for an approximate determination of the solid constituents of the urine. To this
end ascertain the specific gravity and multiply the two last
decimal places by 2 (Trapp), or 2.33 (Neubauer). For
example, a specimen of urine gave the specific gravity
1.016; then in 1000 grams there would be about 37 grams

of solid matter. Especially important are those cases, where in a small volume we find a low specific gravity, and in a large volume, a high specific gravity.

In pathological urine the albumen and sugar most affect the specific gravity; if the latter be high, and the urine pale, sugar or albumen would be indicated as present.

Usually acute inflammations, meningitis, mellituria, *increase* the specific gravity, while it is *lowered* by chronic troubles, hydræmia and kidney affections.

14. Normal urine is acid, but can acquire a transitory alkalinity through the ingestion of alkaline carbonates, and alkaline salts of vegetable acids. Acid urine has some importance for the practitioner, as it favors the formation of certain sediments and concretions and causes an irritation of the kidneys and urinary passages (Vogel). The degree of acidity of urine increases rapidly in rheumatism, pneumonia and pleuritis. The alkalinity of pathological urine should also be carefully noticed. If it originate from potassium carbonate, it would be one of the most unfavorable precursors of brain trouble. Arising from ammonium carbonate, uræmia (the urine often brown colored from admixture of hæmatin), or catarrh of the bladder (in this case, mostly cloudy, from mucus and pus), would very probably be indicated.

III.

THE MOST IMPORTANT NORMAL CONSTITUENTS: THEIR OCCURRENCE IN NORMAL AND PATHOLOGICAL URINE AND THE CHEMICAL DETECTION OF THE SAME.

15. The normal constituents never occur in any constant ratio. Their quantity depends:—

(a) On the manner of life, particularly the nourishment

of the respective individual, his bodily constitution, the quality and quantity of nourishment.

(b) Upon the time of day and the activity of the excreting organs.

(c) Upon the pathological changes.

A disturbance of the normal proportion of the urine constituents is in many instances valuable to the practitioner in his diagnosis, inasmuch as it has been observed that in certain diseases there is not only an *increase*, but also a *decrease*, of the constituents regarded as normal. It is, however, necessary that an accurate knowledge of the mode of nourishment, etc., as in a and b, be obtained, and in addition, that frequent chemical examinations be made. As a consequence of the variation of specific gravity, we recognize the fact that the ratio existing between the solids and the water in urine cannot be constant; it changes from about 12 to 60 grams in 1000 grams of urine.

16. The solid constituents and water are determined quantitatively by evaporating a small and weighed quantity of the urine upon a water bath, and drying the residue in an air bath at 100° C. The method is, however, inaccurate, because in the process of drying the acid sodium phosphate exerts a decomposing influence upon the urea. Therefore, we resort to the use of an apparatus intended to catch the ammonia resulting from the decomposition and determine it. Or, to avoid any trouble, we determine at once the quantity of solids by the specific gravity as given § 13.

17. The fixed salts are estimated by evaporating a measured volume of urine to dryness and igniting over a naked flame until the carbonaceous matter has been completely consumed. In doing this care should be taken that (a) the

temperature does not become so great as to cause the volatilization of chlorides, and (*b*) the carbon does not reduce
the sulphates and phosphates. To avoid any such risk it is
advisable before converting the mass entirely into ash, to
exhaust it with hot water, filter and wash filter paper and
carbon remaining on it, and the filtrate with the wash water,
evaporate to dryness, and then heat to a gentle redness
in a weighed covered porcelain, or better, platinum crucible, allow to cool and then weigh. The difference between
the weight of the empty crucible and the second weight
will be the weight of the fixed salts.

18. The quantity of urea occurring in normal urine
varies, depending largely upon the food ingested and the
weight of the individual. A mixed diet usually shows
from 2.5 to 3.2 per cent.

In all inflammatory diseases, especially in acute brain
trouble, in rheumatism, and in dropsy, if diuretics be administered the amount of urea is *increased*. It is *decreased*,
on the other hand, by neuralgic processes, chronic diseases,
wherever a change of the substance underlies the affection,
in diseases of the spinal cord and kidneys. In typhus, at
first there is an increase of urea, but it rapidly falls, while
it rises in meningitis and remains almost constant in quantity.

Qualitative Detection and Quantitative Estimation of Urea.

19. 20 to 25 c.c. of urine are evaporated to a syrupy
consistence, upon a water bath, the residue repeatedly exhausted with alcohol, filtered and the alcohol expelled by
evaporation upon a water bath. Urea remains behind
somewhat discolored. (Plate I, Fig. 1.) If it be now dissolved in a small quantity of water, and oxalic or nitric

acid added, combinations of urea with these acids will separate in white shining leaflets or hexagonal plates. (Plate
I, Fig. 2.) When the urea is present in minute quantity
the urine is mixed with nitric acid and the formation of
crystals observed under the microscope. Musculus (Pharm.
Centralblatt 15, 161) detects urea in solution by means
of a paper upon which there is a urine ferment. The
latter is prepared by filtering ammoniacal urine through
filter paper, washing the filter, drying at 35–40° C., and
finally the paper is colored with turmeric, again dried and
preserved in closed glass vessels. This paper retains its
sensitiveness for some time. To detect urea immerse it in
a *neutral* urine, and in the presence of the former it will be
decomposed by the ferment into ammonium carbonate and
the paper rapidly becomes brown in various places.

20. Various quantitative methods for the determination
of this constituent have been proposed. That of Liebig
seems to be most generally employed, and yields excellent
results. On adding a dilute mercuric nitrate solution to a
dilute urea solution, and neutralizing the free acid gradually with sodium carbonate, a voluminous, flocculent precipitate will form. Continuing this alternating addition of
mercuric nitrate and sodium carbonate a moment will occur
when the solution of mercuric nitrate added will produce,
with the sodium carbonate, a yellow coloration of mercuric
oxide or basic mercuric nitrate. The solution will then no
longer contain any free urea, but this will be in combination
with mercuric oxide, two equivalents of the latter, 2 HgO
$= 432$, to one equivalent of urea $CO(NH_2)_2 = 60$, forming
$2 \, HgO \, ; \, CO(NH_2)_2$. For convenience we use a solution of
HgO in nitric acid, each cubic centimetre of which will
equal 0.010, or ten milligrams of urea. The reaction

occurs between one molecule of urea and two molecules of mercuric oxide; and to prepare a standard solution we follow the equation—

$$60 : 432 \quad :: 10 \qquad\qquad : x = 72.$$

$CO(NH_2)_2 : 2HgO :: 10$ grms. urea : $x = 72$ grms. the quantity of mercuric oxide to be dissolved in a porcelain dish on a water bath in strong nitric acid, and diluted with distilled water to 1.000 cubic centimetres. But experiment has shown that 5.2 grams HgO should be added, to allow for action upon the indicator—sodium carbonate —leaving 5.2 milligrams HgO in excess in each cubic centimetre of the mercury solution over and above the required quantity, to unite with the urea. Therefore, dissolve 77.2 grams HgO,* in strong nitric acid, evaporate excess of latter on a water bath until the liquid becomes of a syrupy consistence. Treat the residue with water, and dilute to 900 cubic centimetres.† Knowing the approximate strength of the latter solution, we determine its exact titre by means of a normal urea solution, prepared by dissolving two grams carefully dried urea in a little water, and diluting to exactly 100 cubic centimetres with distilled water. Then of this solution,

100 c.c. = 2 grams urea.

10 c.c. = .200 milligram urea.

Having done this, we remove 10 c.c. of the urea solution to a beaker, and, by means of a burette, gradually add the mercuric nitrate solution, mentioned above, until a drop of the liquid brought in contact, by means of a glass rod, with

* Prepared according to Dragendorff, by the precipitation of a solution of 96.855 grams pure mercuric chloride by dilute sodium hydrate. Wash and dry.

† In case any basic nitrate of mercury should separate on dilution with water, allow it to settle, pour off the supernatant liquid, and dissolve the precipitate in a few drops of strong nitric acid, and then add to the original liquid.

a drop of a saturated solution of sodium carbonate, yields a yellow precipitate. Note the exact number of cubic centimetres of mercuric nitrate used. If the latter solution had been exactly standardized, just 20 cubic centimetres would be required for the 10 c.c. of the urea solution. The number, however, of cubic centimetres of mercuric nitrate solution will be less than 20 c.c. Then, in order to bring it up to the proper titre, we make the following dilution : *e. g.*, if 18.5 c.c. of the approximate mercuric nitrate solution were necessary to precipitate 10 c.c. urea solution, 1.5 c.c. of distilled water must be added for every 18.5 c.c. of the original solution, or 15 c.c. for every 185 c.c. of the original approximate mercuric nitrate solution. As we had at first 900 c.c., and removed 18.5 for experiment, there remained 881.5 c.c.; then, as the dilution for every 185 c.c. of mercuric solution is 15 c.c., the corresponding dilution for 881.5 c.c. would be as many times 15 c.c. as 185 c.c. are contained in 881.5 c.c., or 4.76×15 c.c.= 71.40 c.c. distilled water, which, when added to the mercuric solution, will bring it up to the proper strength.

In this method of determining urea in urine, by means of mercuric nitrate, it is necessary to remove the phosphoric and sulphuric acids from the urine, which is accomplished by means of a *barium mixture* (1 part of a cold saturated solution of barium nitrate, and 2 parts of a cold saturated barium hydrate solution).

Execution of the Method.

Measure off a definite volume, say 40 c.c. of urine, into a beaker glass, add half its volume of the barium mixture, then filter through a dry filter, and take 15 c.c. from filtrate. These 15 c.c. would contain 10 c.c. of urine (because the latter had been diluted to half its volume by the barium mixture). Now fill a Mohr's burette to the zero mark with the standard mercuric nitrate solution, and permit the same to run into this urine mixture, drop by drop, until an increase in the precipitate can be no longer noticed. Take out a drop from the well-stirred solution, by means of a glass rod, place it upon a watch glass and bring a drop of the sodium carbonate solution in contact with it. If the mixture remains white, continue the addition of the mercuric solution to the urine, and repeat the test. In this way proceed until the sodium carbonate solution causes a distinct yellow colored precipitate. The number of cubic centimetres of the mercuric solution used multiplied by .010 gram will give the number of milligrams of urea contained in the 10 c.c. of urine. This, multiplied by 10, will give the quantity in 100 parts, or the percentage.

21. Errors that belong to this method and the corrections for the same are—

(a) Corrections for volume of reagent required.

In standardizing the reagent the proportion by volume was 20 c.c. (= 2 vols.) of the reagent to 10 c.c. (1 vol.) of the pure urea solution, and as each c.c. of the reagent contained in excess of that actually required to precipitate the urea present 5.2 mgrms. HgO as nitrate, to react upon the indicator, the 20 c.c. of reagent employed contained $5.2 \times 20 = 104$ mgrms. unprecipitated HgO which were finally distributed through 30 c.c. of liquid. Hence $104 \div 30 = 3.47$ mgrms. of HgO present in each c.c. of the final mixture.

Obviously, a similar proportion will exist when 10 c.c. of urine containing 3 per cent. urea are mixed with 5 c.c.

barium mixture, and then 30 c.c. of the mercuric nitrate solution added.

But, when the undiluted urine contains over 3 per cent. urea, there will be required for 15 c.c. of the urine mixture more than 30 c.c. of the reagent, and consequently the excess of HgO present will be under a *less* degree of dilution than existed in standardizing the reagent, and therefore the final reading would be a little too low.

This discrepancy in regard to dilution, when over 30 c.c. of the reagent are required for 15 c.c. of the urine mixture, may be corrected by adding to the urine mixture ½ c.c. of distilled water for each c.c. of the reagent employed above 30, and then repeating the titration.

Thus, if 40 c.c. of the reagent are employed for the first titration, we add to 15 c.c. of the urine mixture 5 c.c. of water, and then repeat the titration.

So, on the other hand, if *less* than 30 c.c. of the reagent are required for 15 c.c. of the urine mixture, the excess of HgO present in the reagent will be under a *greater* degree of dilution than was present when the reagent was standardized. This difference in conditions may be compensated for by deducting .1 c.c. for every 4 c.c. of the reagent required less than 30.

Thus, if 22 c.c. are required—or 8 less than 30—then 22 — .2 = 21.8 c.c. the quantity of reagent actually required for the precipitation of the urea present, and still leave in solution the same relative proportion of HgO to act upon the indicator as was present when the reagent was standardized.

(*b*) In the sodium chloride present. Either remove the chlorine with silver nitrate, or if the quantity of sodium chloride does not exceed 1 to 1½ per cent., it is only necessary, in order to obtain the approximate number of milligrams of urea in 10 c.c. of urine to deduct 2 c.c. from the number of cubic centimetres of mercuric nitrate required in the estimation.

(c) When the urine contains albumen remove it before the estimation is made, by coagulation and filtration.

(d) When ammonium carbonate is present add the barium mixture, and expel the ammonia by boiling. To estimate the ammonia titrate the urine with a normal sulphuric acid solution.

Salkowski (Zeitschrift für physiologische Chemie, 4, 80.) asserts that Liebig's method for the estimation of urea by means of mercuric nitrate does not yield the quantity of urea, but the approximate quantity of nitrogen in the urine. From his experiments it appears that in the presence of amido and uramido-acids the method furnishes not the urea alone, but the entire quantity of nitrogen in the liquid.

Fowler's Method for the Estimation of Urea.

22. Determine the specific gravity of the urine, and also that of a solution of sodium hypochlorite intended to decompose the urea, then to one volume of the urine add seven volumes of hypochlorite solution, multiply the specific gravity of the hypochlorite solution by 7, and add the result to the specific gravity of the urine. Divide the result of the addition by 8, in order to obtain the mean specific gravity of the mixture, and in the course of two or three hours again determine the specific gravity of the mixture. Deduct this last specific gravity from the mean specific gravity, and multiply the result by .77, and the product will be the percentage of urea. Care must be observed,

C

in the taking of each specific gravity, that the temperatures of the liquids be the same.

Example :—

Sp. grav. of urine $= 1030 \times 1$ vol. $\doteq 1030$
Sp. gr. of hypochlorite $= 1027 \times 7$ vols. $= 7189$

$$8)8219$$

mean sp. grav. $= 1027$
and after decomposition the sp. grav. $= 1024$

$$3 \times .77 =$$

2.31 per cent urea.

The Hypobromite Method for the Estimation of Urea.

23. This method is based on the fact that when urea is exposed to the action of a hypobromite, decomposition ensues, resulting in the formation of an alkaline bromide, carbon dioxide and nitrogen gas. The latter is collected, and its volume measured.

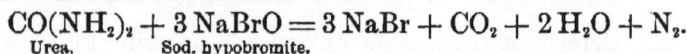

$$CO(NH_2)_2 + 3\,NaBrO = 3\,NaBr + CO_2 + 2\,H_2O + N_2.$$
Urea.   Sod. hypobromite.

Preparation of the Sodium Hypobromite Solution.—The directions of Knop should be followed in preparing the solution, i. e., dissolve 100 grams sodium hydrate in 250 c.c. water, allow to cool, and mix with it 25 c.c. bromine.

In making sodium hypobromite two molecules of sodium hydrate are required for two atoms of bromine. Knowing the density of the latter (about three) we can easily calculate the approximate quantity of sodium hydrate necessary to form hypobromite with the 25 c.c. bromine. Multiply the volume (25 c.c.) bromine $\times$ 3 (density) $= 75$ weight of bromine. To ascertain how much sodium hydrate will be required by the bromine we employ the following equation: 160 :   80   :: 75 : x

Br_2 : 2 NaHO :: 75 : x = 37.5 grams, the quantity of sodium hydrate required by the 75 grams of bromine, and 100 grams NaHO — 37.5 = 62.5 grams, the excess of sodium hydrate which will absorb the liberated CO_2 evolved from the urea in the practical use of the reagent.

Execution of the Method.

Hüffner (Journ. f. prakt. Chemie, Neue. F. Bd. 3, p. 1) employs the following apparatus. The vessel ·c, of about 100 c.c. capacity, is in intimate combination with a, of 10–12 c.c. capacity. They are connected by means of a tolerably wide neck (1.5 centimetres diameter). Between them is b, an air-tight glass stop-cock, the aperture of which is not more than 7–8 millimetres wide. The upper contracted portion d fits closely, by means of rubber, the neck of the upper part of the flask that has been prepared for the purpose. In this manner there is formed a dish, k, of from 4–5 centimetres depth, in the middle of which the contracted portion d projects about 1 centimetre and extends at the same time into the eudiometer e, which is about 30 centimetres long and 2 centimetres wide, divided into ⅓ cubic centimetre, and accurately graduated. The arms f of the iron stand render the apparatus secure. The lower arm clasps the vessel c immediately above the cock b, while the upper arm holds e firmly in position.

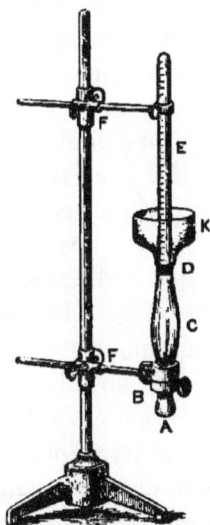

The urea is determined in this apparatus as follows: Aided by a long-necked funnel, fill a and the aperture of the stop-cock with the urine, and close the stop-cock. Then pour equal volumes of the hypobromite solution and distilled water into c, filling it up to the edge. In k pour a saturated sodium chloride solution, making a layer 2 centimetres high, which will serve as a bar to the escape of any gas.

During this time a few air bubbles will be liberated from c. When they have disappeared, invert over d the eudiometer e, filled with water, and when this has been fastened the preparations cease. With one turn completely open the cock b and bring in sudden contact the two solutions. Owing to its higher specific gravity the hypobromite solution will sink, mix with the urea solution and induce the decomposition of the latter with lively evolution of nitrogen gas.

Not more than two or three minutes will elapse from the time of the opening of the stop-cock b and the cessation of the rapid gas liberation, if the hypobromite solution is concentrated and freshly prepared, and the first contact and mixture of the solutions has been sufficiently rapid and complete. The eudiometer, after standing a while, is carefully removed from c, and the volume of nitrogen measured over water, as in Dumas' nitrogen estimation. 1 gram of urea, according to its formula, yields 370 c.c. nitrogen at 0° and 760 mm. pressure. In calculating the result, use the following formula:—

$$p = \frac{100 \ v \ (b - b')}{760. \quad 370. \ a \ (1 + 0. \ 003665 \ t)} \text{ in which}$$

p represents the weight of the urea for 100 c.c. urine.
a represents the volume of urine used.

v the volume of nitrogen read off.

b the barometric pressure.

t the observed temperature during the measurement of nitrogen.

b' tension of vapor of water for this temperature (see Table for Tension of Vapor of Water).

The urine should be diluted three to four times its volume for this method.

24. The frontispiece represents another very simple and convenient form of apparatus, which can be employed in the estimation of urea. It consists of a bottle A, containing a test tube B, and a large glass cylinder C, in which is suspended a graduated burette. The latter is connected by means of a rubber tube with A. In making an analysis with this apparatus, introduce about 5 c.c. of the urine into the test tube B, while about 15 c.c. of the hypobromite solution are brought into A, exercising care not to bring the liquids in contact with each other. The graduated burette is now lowered in C, until the zero mark is on a level with the surface of the water in the cylinder, and the connection between the burette and the bottle accurately made. A is then so inclined that the urine in B will drop into the hypobromite solution. Decomposition at once occurs, accompanied by effervescence. Gradually raise the burette as the nitrogen is evolved, and when the reaction ceases, shake the vessel A and allow to stand for a few minutes until it acquires the temperature of the room in which the operation was performed. The water within and without the burette is leveled, and the cubic centimetres of nitrogen gas read off. This number (say 10 c.c.) multiplied by .027 would represent in grams the quantity of urea in 5 c.c. urine.

For the bottle A can be substituted the apparatus D.

In its arm *b* introduce with the aid of a pipette a given volume of urine, and in *c* place the solution of hypobromite. The connection with the graduated burette is made as before. When ready, carefully remove D from the clamp, and with the hand slightly incline the vessel, permitting the urine to pass drop by drop into the hypobromite solution. It is believed that this careful addition of the urine to the decomposing agent ensures its complete breaking up. The further manipulations are the same as those already described. This piece of apparatus was devised by Dr. Williams, of Boston.

25. It is generally admitted that under the action of the hypobromite reagent, and also under that of an alkaline hypochlorite, only about 92 per cent. of the total nitrogen of the urea is evolved in its free state. M. Mehu, in 1879, proposed to remedy this defect by mixing cane or grape sugar with the urine, before the addition of the reagent.

But, quite recently, Professor Wormley has shown that, under certain conditions, the whole of the nitrogen is set free by the reagent, even without the addition of sugar. These conditions, according to this observer, are the following:—

(1) The reagent should be freshly prepared.

(2) The urea solution should be wholly added to the reagent, none of the reagent being allowed to mix with the urea solution in the containing bulb or tube.

(3) The amount of urea operated upon should not exceed over one part to about 1200 parts of the somewhat diluted reagent.

It is also important that the urea solution be added to the reagent in small portions at a time, thoroughly mixed, and the effervescence allowed to cease before any further addition.

According to Cotton (Chem. Centralblatt, 1875, p. 263), the decomposition of urea by sodium hypobromite is hindered by certain antiseptics, as sulphurous acid, sulphites, hyposulphites, iodine, carbolic acid, etc. ; delayed by such as chloral, and hastened by peroxides, acid potassium chromate, etc.

Musculus' Method for the Estimation of Urea.

26. Musculus (Archiv. der Physiologie 12, 214), in his investigations on urine ferment, remarks that the best material for the preparation of the latter is the thick, mucous, ammoniacal urine of persons suffering with catarrh of the bladder. On adding alcohol to such urine the mucin is coagulated to a film-like mass, and can be easily separated from the liquid. The precipitate is dried at a gentle heat, pulverized and kept in closed glass vessels.

This ferment is excellently adapted to the quantitative estimation of urea.

10 c.c. of urine mixed with a small quantity of sodium carbonate, then diluted 10 times with water, are colored with a few drops of litmus, accurately neutralized by a dilute acid, 0.2 grams ferment powder added and warmed upon a water bath to 35–40° C. In an hour the urea is completely decomposed. By titration with normal sulphuric acid the amount of ammonia formed is determined, and from this the urea calculated. Creatin and creatinin are not decomposed by the ferment.

Uric Acid.

27. The Uric acid found in urine is partly combined and partly uncombined; its quantity ranges from 0.2 gram to 1 gram in 24 hours. Disturbed digestion, fevers, affections

of the respiratory organs and disturbance of the blood circulation *increase* the quantity of uric acid, while in its decrease it is analogous to urea, and like the latter may be converted into ammonium carbonate.

28. For its detection evaporate, on a water bath, 100 to 200 c.c. of urine, from which any albumen present has been previously removed by coagulation and filtration, to a syrupy consistence. Dissolve out the urea and extractive matters with alcohol, and the residue will consist of uric acid, mucin and fixed salts.   ·

Add a little nitric acid to a portion of the residue and warm, when nearly all will dissolve. On careful evaporation on a water bath there will remain a red-colored residue, which moistened with ammonium hydrate (avoid an excess) will assume a purplish-red color—murexide. With a drop of sodium or potassium hydrate this becomes purplish-blue.

Another portion of the first residue dissolved in potassium hydrate, then mixed with hydrochloric acid and allowed to stand for some time, will yield crystals of uric acid. (Plate I, Fig. 4.)

When much uric acid is present add hydrochloric acid to 200–300 c.c. of urine, and allow the same to stand 24 hours. In that time the uric acid will have separated out in colored crystals, and can be readily recognized under the microscope (See Plate I, Fig. 5).

29. In estimating it quantitatively we pursue essentially the directions in the preceding section, viz : Mix from 200 – 300 c.c. urine with hydrochloric acid (3 – 4 c.c.), and allow to stand for 24–48 hours; the temperature being as low as possible. The separated crystals of uric acid are collected on a previously washed, dried and weighed

PLATE I.

FIG. 1.

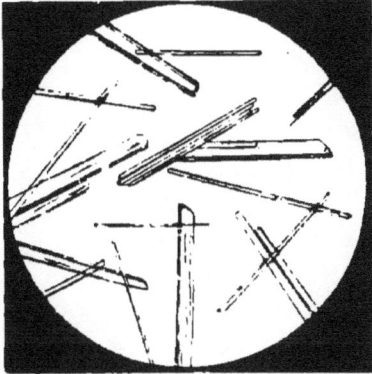

Pure urea from an alcoholic solution.

FIG. 2.

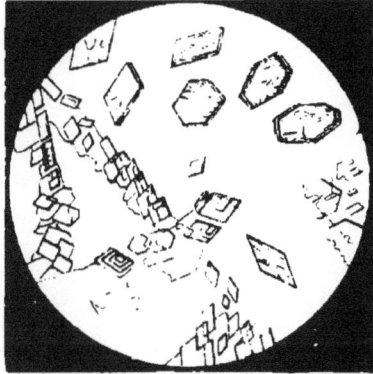

Urea Oxalate (upper half); urea nitrate (lower half.)

FIG. 3.

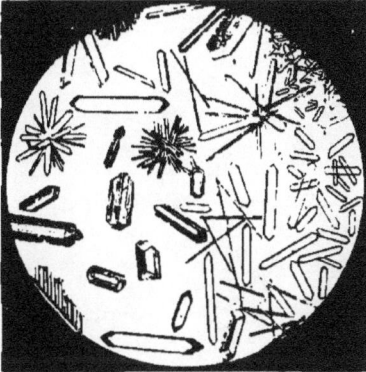

Hippuric Acid from normal human urine.

FIG. 4.

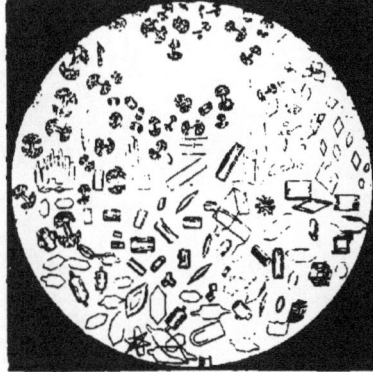

Various forms of Uric acid from urinary sediments.

FIG. 5.

Uric acid.

FIG. 6.

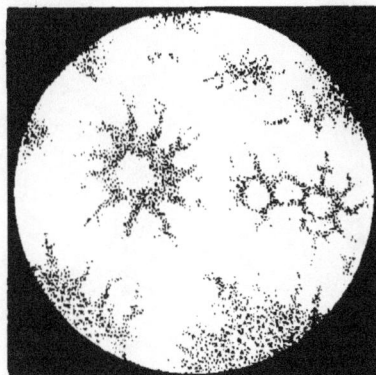

Natural Sodium Urate.

filter paper, washed well with water and after drying, weighed. The first weight of the filter paper subtracted from the last weight will give the amount of uric acid in the quantity of urine employed.

Salkowski (Virchow's Archiv, 68, 1), proposes the following method for the determination of uric acid : 200 c.c. urine are rendered strongly alkaline with sodium carbonate (10 c.c. of concentrated solution); after an hour 20 c.c. of a concentrated ammonium chloride solution are added, and the whole allowed to stand at a low temperature for 48 hours, then filtered through a weighed filter and washed two or three times with water. The filter is then filled with dilute hydrochloric acid (1 part commercial acid to 10 parts water), and the filtrate preserved. The addition of hydrochloric acid to the precipitate on the filter is repeated several times, until all the ammonium urate has been converted into uric acid. Let the filtrate stand six hours ; the uric acid that separates from it in this time is brought upon the same filter; wash the precipitate twice with water, then with alcohol, until the acid reaction of the filtrate passing through disappears, and dry at 110° C., and weigh. To the number found add 0.030. Dilute urine should be evaporated until its specific gravity becomes 1.017–1.020.

Oxaluric and Hyposulphurous Acids.

30. Recently Schunk discovered oxaluric acid in normal urine, existing there in combination with ammonia. It is a white, acid tasting, crystalline powder, difficultly soluble in water. The ammonium salt is soluble in water.

By dissolving uric acid in warm, very dilute nitric acid, and adding ammonium hydrate just as the solution is cold, then evaporating to crystallization, we can easily obtain crystals of ammonium oxalurate. Hydrochloric acid separates the free oxaluric acid as a white powder from concentrated solutions of the ammonium salt. The acid dissolved in water and recrystallized forms beautiful aggregations or rosettes.

A. Strumpell (Archiv. d. Heilkunde, 17, 390), has discovered hypo-sulphurous acid in the urine of a typhoid patient. The acid can be estimated quantitatively by precipitating with barium chloride and in the filtrate from the barium sulphate (which will contain barium hyposulphite in solution), the barium hyposulphite can be oxidized by means of a few drops of nitric acid, and the S_2O_2 can be calculated from the amount of barium sulphate formed.

The Chlorides in Urine.

31. These occur principally as sodium chloride. They average in 24 hours about 15 grams in 1600–1700 c.c. urine. In a healthy, robust man they can become even more abundant.

The *decrease* of chlorides is of particular interest to the diagnostician, and has been noticed :—

(*a*) in all cases where the chlorides have not been reabsorbed, as in cholera, certain stages of typhus, inanition following pathological changes, etc.

(*b*) in abnormal transudations.

(*c*) in acute exudations in the following pathological processes ; pneumonia, pleuritis, peritonitis, pericarditis, endocarditis, meningitis, typhus, acute miliary tuberculosis, and the like. The disappearance of the chlorides in rheumatism of the joints and pericarditis is characteristic. Their quantity decreases so rapidly then that by comparison of the tests made within a few hours of each other we can determine upon any conspicuous change in the course of the ailment.

The qualitative test for the detection of the chlorides consists in acidifying the urine with nitric acid, then adding silver nitrate, when chlorine, if present, will be precipitated as silver chloride.

32. Quantitatively the chlorides can be estimated gravimetrically or volumetrically.

(a.) A definite volume (say 10 c.c.) of the urine is evaporated to dryness on a water bath with a few drops of nitric acid and about two grams potassium nitrate. It is then ignited over a naked flame until all the carbonaceous matter has been destroyed, allowed to cool, dissolved in water acidified with nitric acid, heated to almost the boiling point, when silver nitrate is added and the solution stirred with a glass rod. This will cause the silver chloride to collect and settle. The addition of a drop of silver nitrate to the supernatant liquid will show whether the precipitation is complete. If so, filter the solution and bring the silver chloride upon a filter, wash rapidly with hot water, dry, then separate the precipitate as fully as possible from the filter and place it in a weighed porcelain crucible. The filter is reduced to ash on the inverted lid of the crucible. The traces of silver chloride which are reduced to the metallic state can be reconverted into chloride by moistening the ash with a drop of nitric acid and then a drop of hydrochloric acid. Heat carefully and evaporate excess of acid, allow to cool, place the lid upon the crucible, to which apply a low heat, then, after cooling, weigh.

To calculate the quantity of sodium chloride the following equation is employed:—

$$AgCl \; : \; NaCl \; :: \; \text{wt. of prec.} \; : \; x \; = \; \text{amt. of } NaCl$$
$$143.5 \; : \; 58.5$$

in the 10 c.c. urine; multiply x by 10 and the percentage will be obtained.

To calculate the quantity of chlorine change the second term of the equation to 35.5 and x will equal the quantity of that element in a given quantity of the urine.

(b.) Of the volumetric methods there are several.

I. That of Liebig is based on the circumstance that sodium chloride acting upon mercuric nitrate causes the formation of the soluble compounds, mercuric chloride and sodium nitrate, and that so long as there is a chloride present in the urine a precipitation of the urea cannot occur, but just as soon as all the chlorine has entered into combination with the mercury and the mercuric nitrate is added in more than sufficient quantity to cause the preceding change with the chloride, a white cloudiness will appear, resulting from the union of the excess of mercuric oxide with urea. This latter substance then acts as the indicator, and on this behavior the method is founded. As one equivalent of mercuric oxide is equal to two of sodium chloride, the solution of mercuric nitrate is so prepared that 1 c.c. of the latter will equal ten milligrams of sodium chloride. That is, we would make the following calculation

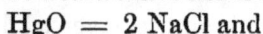

$$HgO = 2 \ NaCl \text{ and}$$

117   : 216   :: 10 : x

2 NaCl : HgO :: 10 grams NaCl : x = 18.461 grams of mercuric oxide, which are to be dissolved in a porcelain dish, on a water bath, with strong nitric acid, and treated in the same manner as described in the preparation of the mercuric nitrate solution for the estimation of urea, (page 21). After dilution it can then be standardized by means of a standard solution of sodium chloride, prepared by dissolving one gram perfectly dried sodium chloride in 100

c.c. of water.   10 c.c. of the latter solution are measured out into a beaker, a small pinch of urea dissolved in it and the mercuric solution added until the appearance of a permanent cloudiness.   The quantity used is then read off.   If, for example, 8.2 c.c. mercuric solution were required, then for every 8.2 c.c. of mercuric solution on hand add 1.8 c.c. distilled water.   A new titration can then be made and the solution will be found up to the proper strength.   10 c.c. should equal 10 c.c. of the sodium chloride solution.   1 c.c. of the mercuric solution will then be equal to 10 milligrams sodium chloride or 0.00606 milligram chlorine.*

In the practical execution of this method the phosphoric and sulphuric acids must first be removed from the urine, which is accomplished by the use of the barium mixture, as given in the determination of urea.   Take 20 c.c. of urine, mix with 10 c.c. barium mixture, and filter through a dry filter.   The solution will be alkaline; and from the filtrate take 15 c.c. (of which 10 c.c. are urine) and make it neutral, or, at the most, very slightly acid, with nitric acid, and then commence the addition of the mercuric nitrate, drop by drop, from a burette.   The first drop of the latter will cause a turbidity, which disappears on stirring the liquid.   Continue adding the mercuric solution until a permanent turbidity is produced; read off the number of cubic centimetres of the mercuric solution used and multiply these by .010, and the product will represent the number of milligrams of sodium chloride contained in the 10 c.c. of urine.   This product multiplied by 10 will give the percentage of sodium chloride.

II. Neubauer's modification of Mohr's method.

10 c.c. urine are brought into a platinum or porcelain

* $NaCl \quad Cl \quad NaCl$
$58.5 : 35.5 :: .010 : x = 0.00606$

dish, 2 grams powdered potassium nitrate, free of chlorine, added, and the whole evaporated to dryness on a water bath or hot plate. The residue is heated gently at first, over a naked flame, more intensely later, until the carbonaceous matter is completely oxidized, and the fused mass, upon cooling, is perfectly white in appearance. In cooling, withdraw the flame slowly, so as to prevent any likelihood of cracking the porcelain dish or spurting of the fused substance. The residue of salts is now dissolved in about 30 c.c. of water, washed into a beaker, the platinum or porcelain dish carefully washed out, and the wash water added to the solution of the salts, and the whole evaporated down to about 30 c.c. The solution will be alkaline, from the decomposition of the potassium nitrate. Dilute nitric acid (or acetic, in which case neutralization with calcium carbonate is unnecessary, as this acid does not decompose silver chromate) is added, drop by drop, to the liquid, until the latter yields a faint acid reaction, which is removed by the addition of a small quantity of precipitated calcium carbonate. The latter, if added in excess, need not be filtered off. To the solution thus prepared add 2 to 3 drops of a cold saturated solution of neutral potassium chromate, $K_2 CrO_4$, which acts as the indicator. Silver has a greater affinity for chlorine than for chromic acid; therefore, no combination will take place between the silver and chromic acid until all the chlorine has been satisfied. The standard silver nitrate solution is now allowed to run into the sodium chloride solution, drop by drop, from a burette, with constant stirring, until a distinct orange color is produced, which remains permanent*. The number of c.c.

* *Correction.*—On account of the dilution of the mixture if more than 10 c.c. of the silver solution be required to produce the orange coloration, $\frac{1}{10}$ of a c.c. is deducted from the number of c.c. silver solution for every 5 c.c. used above 10 c.c.

silver solution used, multiplied by .010 gram, will give the number of milligrams of sodium chloride in the 10 c.c. employed. This number multiplied by 10 will furnish the percentage. The amount of chlorine is found by multiplying the number of c.c. silver solution used by 0.00606 gram, and the product multiplied by 10 gives the percentage of chlorine.

Preparation of Silver Nitrate Solution.

The standard silver nitrate solution is prepared as follows: It is made of such strength that 1 c.c. will be equal to 10 milligrams sodium chloride. The reaction takes place between one molecule of silver nitrate and one molecule of sodium chloride, represented by the equation—

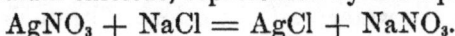

$$AgNO_3 + NaCl = AgCl + NaNO_3.$$

Therefore, in order to determine the quantity of silver nitrate necessary to make a solution of standard strength, we use the following proportion :—

58.5 : 170 :: 10 : x.

$NaCl : AgNO_3 :: 10$ grms. NaCl : 29.059 grms. $AgNO_3$. That is, 29.059 grams chemically pure, fused silver nitrate are dissolved in a little water, and the solution diluted to 1 litre; then

1000 c.c. = 10 grms. NaCl.

1 c.c. = .010 grm. NaCl.

If necessary, the solution can be standardized by means of a standard solution of sodium chloride, made as follows: dissolve one gram thoroughly dried sodium chloride, in a little water, and dilute to 100 c.c. with distilled water; measure off into a beaker 10 c.c. of the solution, add two or three drops of potassium chromate solution, run in from a burette, with constant stirring of the liquid in the beaker, the silver solution, until an orange coloration appears,

which is persistent. 10 c.c. of the silver solution should
have been required to produce this coloration, and if more
or less than this quantity were required, then make the
necessary corrections, as given under Liebig's mercuric nitrate solution for the estimation of urea.

III. Primbram's Method. Primbram has proposed a
slight modification to Neubauer's method. Instead of de-
stroying the organic matter by ignition with an alkaline
nitrate, he adds to a measured volume of urine a few c.c. of
a saturated solution of potassium permanganate, and heats
to almost boiling, when oxide of manganese separates.
The permanganate is added until the liquid, on warming,
retains a purple color, when it is filtered, the precipitate
washed with hot water, and the colored filtrate decolorized
by the addition of a little oxalic acid. Any excess of the
latter is neutralized by a little precipitated calcium carbonate. The solution is now reduced by evaporation to a
definite volume (say 10 or 20 c.c., or the original volume
employed), and titred with the silver solution, as in Mohr's
method.

IV. Falck (Berichte d. deutsch. Chem. Gesellschaft, 8,
12) recommends the following in estimating chlorides in
urine : after the evaporation of 10 c.c. urine, and ignition of
the residue with potassium nitrate, the salts remaining are
dissolved in a little water, and washed into a beaker glass ;
the alkaline solution is acidified with nitric acid, and, after
the addition of 4 c.c. ammonium ferric sulphate solution,
made blood-red by the aid of 1 or 2 drops of a titrated
ammonium sulphocyanide solution. The standardized silver
nitrate solution is now added from a burette, until the red
coloration just disappears. The number of cubic centimetres of the latter solution thus required do not exactly

correspond to the chlorine in the liquid, because, in the incineration with potassium nitrate, nitrites are invariably produced, and the nitrous acid liberated upon the addition of nitric acid affects the final reaction. Therefore, again ignite 10 c.c. urine, strongly acidify the solution with nitric acid, mix the solution with an excess of silver nitrate solution, so that all the chlorine present will be in combination with the silver. The solution is now warmed upon a water bath, to expel the nitrous acid, then cooled, mixed with 5 c.c. iron alum solution and the ammonium sulphocyanide added, drop by drop, until the red coloration of iron sulphocyanide no longer disappears. The difference between the required number of cubic centimetres of silver and sulphocyanide solutions represents the chlorine contained in the urine.

The following solutions are necessary in the above method : (a) Solution of silver nitrate, of which 1 c.c. corresponds to 10 milligrams sodium chloride.

(b) Solution of ammonium sulphocyanide accurately standardized with the silver solution, so that 10 c.c. of the former will be required to precipitate the silver in 10 c.c. of the standard silver solution as silver sulphocyanide.

(c) A cold saturated solution of crystallized ammonium ferric sulphate free from chlorine.

PHOSPHORIC ACID.

33. The phosphoric acid in urine exists partly combined with sodium, as acid sodium phosphate, and partly in combination with calcium and magnesium, as calcium and magnesium phosphates. Regarding the increase or decrease of phosphates in pathological changes the following may be observed :—

(a) In the urine of persons suffering from inflammatory

D

diseases, *e. g.*, acute brain affections, acute spinal troubles, the alkaline phosphates are *increased*. They decrease in neurosis, chronic spinal affections, and kidney diseases.

(*b*) The phosphates of the alkaline earths (earthy phosphates) are increased by *meningitis*, especially in acute brain affections and rheumatism. They decrease in kidney and spinal affections, and in neuralgia.

Detection and Quantitative Estimation of Phosphoric Acid.

34. On adding ammonium hydrate in excess to urine the phosphates of calcium and magnesium are precipitated ; the latter as triple phosphate. The phosphoric acid that yet remains in solution, after adding ammonium hydrate, is recognized by acidifying the solution with acetic acid, and then adding a little ferric chloride, when a yellowish-white precipitate of ferric phosphate is produced.

35. Phosphoric acid is best determined quantitatively, volumetrically, by means of a standard uranium acetate solution. The method is based on the insolubility of uranium phosphate in acetic acid. The merest trace in excess of uranium acetate is recognized by the reddish-brown color formed when a drop of the liquid is brought in contact with ferrocyanide of potassium.

The uranium acetate solution is so standardized that 1 c.c. of it equals 0.005 gram of phosphoric acid.

The formula of the precipitate formed by the addition of the uranium acetate to a solution of a phosphate is $2UrO_3$: $P_2O_5 + Aq$. Two molecules, UrO_3 are required to combine with one molecule, P_2O_5, therefore, in the preparation of the standard uranium acetate solution we use the following proportions :—

142 : 576 : : 5 : x

P_2O_5 : $2UrO_3$: : 5 grams P_2O_5 : x = 20.28 grams UrO_3, necessary to combine with 5 grams P_2O_5; then to find the quantity of uranium acetate equivalent to 20.28 grams UrO_3 we make another proportion:—

288 : 442 : : ˙20.28 : x

UrO_3 : $UrO_3 (C_2H_3O_2)_2 + 2H_2O$: : 20.28 grams UrO_3 : x = 31.1 grams uranium acetate, to be dissolved in 900 cubic centimetres of water, about 5 c.c. strong acetic acid added and allowed to stand for a few hours, in order that a precipitate which usually forms may subside. The solution is then filtered and titrated by means of a standard phosphoric acid solution, and diluted after the plan used in standardizing the mercuric nitrate solution for the estimation of urea (page 22).

If uranium nitrate be preferred in the preparation of the uranium solution, substitute in the second equation above, the molecular weight of uranium nitrate ($UrO_3N_2O_5 + 6H_2O = 504$) for that of uranium acetate $UrO_3 (C_2H_3O_2)_2 + 2H_2O = 442$), and the result will be the number of grams uranium nitrate required.

Instead of ascertaining the amount of uranium acetate or nitrate by the two equations above mentioned, we can immediately determine the required quantity of the respective compounds by the following single equations:—

142 : 884 : : 5 : x

P_2O_5 : $2UrO_3(C_2H_3O_2)_2 + 2H_2O$: : 5 grams P_2O_5: x = 31.1 grams uranium acetate.

142 : 1008 : : 5 : x

P_2O_5 : $2UrO_3N_2O_5 + 6H_2O$: : 5 grams P_2O_5 : x = 35.5 grams uranium nitrate.

The solution of phosphoric acid which is used for

standardizing the uranium solution is prepared as follows:
As phosphoric acid itself cannot be weighed, a stable
weighable compound in which it exists in combination with
a base is used for the purpose. Sodium hydrogen phosphate is the salt usually employed. To obtain one molecule of P_2O_5 we must use two molecules of $Na_2HPO_4 + 12$
H_2O. Then to find the quantity of sodium hydrogen phosphate which shall contain 5 grams phosphoric acid, we use
the following proportion :—

$$142 :   716   : : 5 : x$$
$$P_2O_5 : 2Na_2HPO_4 + 12H_2O : : 5 \text{ grams}   P_2O_5 : x =$$
25.211 grams sodium hydrogen phosphate, which will be
equal to 5 grams P_2O_5. The 25.211 grams well crystallized $Na_2HPO_4 + 12H_2O$ are dissolved in a little water
and the solution diluted to 1 litre. Then of this solution—

$$1000 \text{ c.c.} = 5. \text{ grams } P_2O_5.$$
$$1 \text{ c.c.} = 0.005 \text{ gram } P_2O_5.$$

In standardizing, measure off into a beaker 20 c.c. of
the standard sodium phosphate solution, add 30 c.c. distilled water and 5 c.c. sodium acetate solution (prepared
by dissolving 100 grams crystallized sodium acetate in 900
c.c. water, and adding acetic acid until the volume reaches
1000 c.c.). The mixture is then heated on a water bath,
to a temperature between 90 and 100° C., and the uranium
solution gradually added from a burette, the mixture being
stirred constantly, until a drop of the liquid, removed by
means of a glass rod, produces a reddish-brown color
when brought in contact with some powdered potassium
ferrocyanide or a concentrated solution of the same.
20 c.c. of the uranium solution equal to .100 gram P_2O_5

should be required to unite with the P_2O_5 present and give the reaction with the indicator—potassium ferrocyanide.

In this titration a less number of cubic c.c. than 20 of uranium solution will be used, and then, in order to bring it to the exact strength, that is, that 20 c.c. shall be required, make the dilution as given under the standardizing of the mercuric nitrate solution. For example, if 18.4 c.c. uranium solution had been required, then for every 18.4 c.c. contained in the original volume of uranium solution add 1.6 c.c. distilled water.

In the actual analysis, measure off 50 c.c. urine into a beaker, add 5 c.c. sodium acetate solution and heat upon the water bath. Then slowly add the uranium solution, from a burette, with constant stirring of the mixture, until a drop of the latter, removed with the aid of a glass rod, gives a perceptible reddish-brown coloration when brought in contact with the indicator, potassium ferrocyanide. The number of cubic centimetres of uranium solution used is now read off, and then multiplied by the strength of 1 c.c. = .005 gram, and the result will be the quantity of P_2O_5 in the 50 c.c. urine employed.

36. To estimate the phosphoric acid combined with the alkaline earths, add to a measured quantity of urine (say 200 c.c.) ammonium hydrate in excess, and stand aside for a few hours, collect the precipitate of earthy phosphates on a filter, wash and dissolve in as little acetic acid as possible, dilute the solution with water to 50 c.c., add 5 c.c. acetate of sodium solution, warm on water bath, and titrate with standard uranium acetate solution. The number of cubic centimeters of the uranium solution, multiplied by the strength of 1 c.c. (0.005 gram), will furnish the quantity of P_2O_5 combined as earthy phosphates in 200 c.c. urine.

In this determination care should be taken to avoid an excess of sodium acetate, as it affects the delicacy of the reaction of potassium ferrocyanide.

SULPHURIC ACID.

37. Next in importance to the phosphates are the sulphates, which are qualitatively detected in acidified urine by means of barium chloride. (See § 4.)

The quantitative determination of the sulphuric acid is . based on the insolubility of barium sulphate. The method may be executed gravimetrically or volumetrically. If the latter is preferable, a standard solution of barium chloride should be used; 1 c.c. of this solution should correspond to 0.010 gram of sulphuric acid.

To estimate it gravimetrically add about 20 grams potassium nitrate to 100 c.c. of urine, and evaporate to dryness on a water bath, then incinerate over a naked flame until all the carbonaceous matter has been destroyed. The fused mass is then dissolved in water, acidified with hydrochloric acid, brought to the boiling point and an excess of barium chloride solution added. The precipitated barium sulphate is collected on a filter, washed with hot water, dried, as much of it as possible detached from the filter paper and placed in a weighed crucible, the filter paper incinerated on the end of a platinum wire, held over the crucible, and after cooling, the whole weighed. The following equation will serve for the calculation of the result :—

$$233 \quad : 80 \quad : : \qquad \qquad : x$$
$$BaSO_4 : SO_3 \quad : : \quad \text{wt. of prec.} : x$$

In addition to the sulphates in urine, Baumann (Berichte d. deutsch. Chem. Gesell, 9, 54,) has proven that sulpho-acids are also constantly present. According to this chemist, the phenol, indigo, and brenz-catechin forming substances are found in urine as sulpho-acids. To estimate them when both are present, pursue the following course: Strongly acidify the fresh urine with acetic acid, and add an excess of barium chloride, filter off the precipitate after standing one to two hours, wash first with water, then with warm dilute hydrochloric acid, and finally with water. The filtrate and wash water from the precipitate are then warmed for several hours with an equal volume of hydrochloric acid upon a water bath. The precipitate that separates contains, in addition to an amorphous organic substance, barium sulphate, the sulphuric acid of which did not exist as sulphate in the original urine.

COLORING MATTERS IN URINE.

38. Urine Brown, urophain, *increases* in inflammatory troubles, in disorders of the liver, and in icterus, frequently very markedly *decreased* in neurosis.

Urine Yellow, uroxanthin, is increased in violent functional disturbances in the spinal marrow (forming urrhodin and uroglaucin), *e. g.*, in a sudden fall, sudden fright, etc., in acute kidney affections, and in cholera.

Urine abundant in uroxanthin deposits upon long standing, and during alkaline fermentation, a blue sediment (uroglaucin); hence the so-called blue urine (Cholera morbus Brightii).

Urobilin might also be noticed. Jaffé noticed this in both normal and pathological urine, and also in the bile. The pigment is distinguished by the magnificent fluorescence which it exhibits under certain conditions, and by its characteristic spectrum. The urine of persons suffering with fever is rich in this pigment. The spectroscopic study of it reveals an absorption band between Frauenhofer's lines *b* and *F*. With alkalies it shows a characteristic play of colors.

To detect urobilin in normal urine precipitate 100 to 200 c.c. of the latter with lead acetate, and decompose the washed and dried precipitate with an alcoholic solution of oxalic acid. If the solution does not exhibit any absorption lines, mix it with chloroform, and shake up with water. Upon the addition of ammonium hydrate and zinc chloride, the acid alcoholic liquid will yield an exquisite fluorescence, and show sharp, well defined lines in the spectrum.

APPROXIMATE ESTIMATION OF THE COLORING MATTERS.

39. For this purpose either Neubauer and Vogel's color scale or Heller's urophain reaction is employed. In making the latter we proceed as follows: Pour 2 c.c. of colorless sulphuric acid into a small beaker and let flow into this, from a height of about four inches, two parts urine, in a delicate stream. The urine, when mixed intimately with the sulphuric acid, produces an intense garnet-red coloration, providing the sample was normal urine, *i. e.*, having a specific gravity of 1.020, and the quantity eliminated in twenty-four hours being about 1500 c.c. If there has been an increase in the quantity of coloring matter, the urine mixture will be opaque and black ; if the quantity be less than normal the mixture will appear pale garnet-red and perfectly transparent.

Care must be observed in this experiment, that the urine does not contain any sugar, blood, or biliary coloring matter, as these would indicate an apparent increase of the quantity of urophain.

To perform the test for urophain, pour about 3–4 c.c. of pure concentrated hydrochloric acid into a small beaker, and then drop in, while stirring, from ten to twenty drops of normal urine. Usually the quantity of this coloring matter is so slight under normal conditions that the acidulated urine is of a feeble *yellowish-red* color. When the quantity is large the hydrochloric acid is colored from violet to blue. Frequently 1–2 drops of urine suffice to color 4 c.c. hydrochloric acid blue. If a violet color does not appear in from one to two minutes, the coloring substance has not increased above normal, even if the mixture, after standing from ten to fifteen minutes, assumes a dark reddish-brown color. In icteric urine the bile-coloring matters should be removed with lead acetate and the filtrate employed for this test.

IV.

THE ABNORMAL CONSTITUENTS OF URINE; THEIR OCCURRENCE AND DETECTION.

40. These arise in certain disturbances of the health of an individual, and are partly such substances which pass through the kidneys in consequence of altered transudation relations, while they are constantly present in the blood; or they arise from a metamorphosis of the tissues, and are even formed in the latter, and under normal conditions even further transposed, and under abnormal conditions passing through the blood are eliminated by the kidneys.

ALBUMEN.

41. The conditions under which albumen appears in the

urine are by far more numerous than formerly supposed, when it was believed that from the presence of albumen certain diseases could be diagnosed.

Albumen is found

(a) In general sickness, e. g., pure hydræmia, chlorosis, endemic diseases, dropsy. Further, in disturbance of the circulatory organs, heart troubles, and affections of the liver, when, by a difference of pressure, there ensues an infiltration of albumen.

(b.) In diseases of the uropoetic system, e. g., in the so-called sympathetic kidney diseases, in typhus, peritonitis and violent phlogosis which influence hyperæmia of the kidneys. And in the so-called idiopathic affections of the kidneys:

(1) Albuminuria such as is observed in Bright's disease, in nephritis, neoplasma renis. The so-called Bellinic casts, pus sediment in acid reaction and a small quantity of neoplasms even, always distinguish each of these troubles introducing albumen.

(2) Hæmaturia, which may be partly a hemorrhagic capillary hæmaturia, in which fibrous coagula do not appear; or partly a hemorrhagic vascular hæmaturia, in which a blood clot is found; thirdly, and finally, a serous hæmaturia, where no blood corpuscles, but blood coloring matters are present, together with the albumen. If these occur where the specific gravity is over 1.020 it is a symptom of uræmia.

Urine, red in color, rich in albumen, free of blood corpuscles, and having a specific gravity below 1.020 is supposed to contain blotches or collections of blood cells. When the specific gravity rises above 1.020 the quantity of the blood coloring matter in the urine can only be accounted for by the ammonium carbonate, which extracts the hæmatin and becomes thereby a specific uræmic symptom.

42. Very often in pyuria albumen is discharged regularly. The acid or renal pyuria we find in pyelitis, urethral catarrh, nephritis, etc. The alkaline pyuria shows catarrh of the bladder, in combination with renal pyuria or alone, when, however, it is in the pus stage.

43. Finally, when with the albumen, which passes off with pus in phlogosis of the kidneys, not unfrequently an equal or greater quantity of albumen in the interstitial capillary or vascular hæmaturia is separated in the urine. We designate this stage hæmatopyuria.*

44. We find albumen in urine, in addition, in many fevers, remittent as well as intermittent; also in exanthematous affections (measles, scarlet fever, smallpox), further in affections of the respiratory organs (pneumonia, tuberculosis), and after excesses in eating, and after excitement of the animal passions; also after the inhalation of hydrogen arsenide.

Gerhardt (Wien med. Presse, 1871, p. 1.) has frequently observed peptones in urine free of albumen, either as a forerunner or consequence of ordinary albuminuria. Senator declares that peptones exist in every albuminous urine in slight quantities.

Detection of Albumen.

45. Many difficulties are met with when testing for albumen. The first step should be to ascertain the reaction of the urine. Then, if it be neutral or alkaline, acidulate it slightly with nitric acid, and heat the specimen, in a test tube, to 60 or 80° C. Turbidity follows, and very soon re-

* See Folwarczny's Handbuch d. physiolog. Chemie, Wien, 1863.

sults in the coagulation of the albumen.   Alcohol also produces coagulation.

Heller's test is to take a small beaker or large test tube, bring into it about 10 c.c. of urine, then incline the glass, let half this volume of concentrated nitric acid trickle down the side, and at the point of contact of the two liquids, if albumen is present, there will be produced a band-like, sharply defined white zone.   It is true that in the presence of large quantities of urates in the urine a similar layer is produced, not at the point of contact of the urine and acid, but higher up, and it is not sharply defined below, but is rolled up similar to rising smoke.   The other methods of estimating the albumen by alcohol and tannic acid we will pass over.

Galipe (Pharm. Zeitschrift für Russland, 13, 683) recommends the following test for albumen in urine.   In using it the mistaking of phosphates and urates for albumen is impossible.   Fill a reagent glass one-third with a highly-colored picric acid solution, and drop in two to three drops of the urine under examination.   In the presence of albumen there forms immediately a sharply defined white turbidity.   On warming the liquid the albumen collects into balls, which rise to the surface of the liquid and float there.

Quantitative Estimation of Albumen.

46. The urine is first filtered, and from 20 to 100 c.c. of the filtrate are then taken for the estimation of the albumen (we should never have more than from 0.2 to 0.3 gram coagulated albumen).   Concentrated urine—that is, urine containing a large percentage of albumen, should be diluted with water.   The beaker containing the urine under examination is heated on a water bath for half an hour.

If a flocculent precipitate does not appear, from want of sufficient acidity of the urine, add, by means of a pipette, from one to three drops acetic acid, avoiding an excess. When the coagulation is complete, filter through a previously dried and weighed filter. When the .liquid has passed through, wash the albumen with hot water, until a drop of the filtrate evaporated on platinum foil does not leave a residue. The filter and precipitate are dried on a watch crystal at 100° C., and when cooled weighed. After the deduction of the weight of the watch crystal and filter paper, we have the weight of the albumen. One source of error in this method is that in the coagulation of the albumen it may enclose earthy phosphates, and therefore, after ascertaining the weight of albumen, place the latter in a weighed crucible and ignite, allow to cool, weigh, deduct the weight of the crucible + the earthy phosphates from the first weight (crucible + albumen) and the result will be the exact amount of albumen.

The method of Bornhardt for the estimation of albumen is readily applied and consequently of service to the practitioner. It consists in determining the sp. grav. of the freshly precipitated albumen. A delicate sp. grav. bottle is filled with water and weighed, then the moist albumen introduced and the sp. grav. bottle re-weighed. The albumen being specifically heavier than water (1.314), the sp. grav. bottle would, of course, show an increased weight in the second weighing. The quantity of albumen is found from the following formula:—

$$x = d, \frac{1.314}{0.314,}$$

in which d represents the difference in weight of the specific gravity bottle when filled only with water, and then with water and albumen.

SUGAR IN URINE.

47. Brücke contended that sugar in small quantities was a normal constituent of urine, but this view has not met with general acceptance. It is a constant ingredient, however, of urine in but one disease—diabetes mellitus. Here it is eliminated, frequently in such abundance that the urine possesses a sweet taste, and cloths soaked in it, after the volatilization of the urine, become sticky, and look as if they had been coated with honey. Sugar appears in the urine after injury to the fourth ventricle of the brain; therefore, this was believed to be the cause of the disease in diabetes mellitus, but the connection between the irritation of the brain and the sugar separation is yet perfectly in the dark. Sugar is also found in galactostasis, now and then in dyspepsia, in diseases of the lower extremities and hypochondria, in the convalescent stage of cholera, in Bright's disease, but requires yet, in many cases, further confirmation.

48. The urine in diabetes is usually very pale, of peculiar odor and high sp. grav., 1.030–1.052. Freshly passed, it very rarely gives a strong acid reaction, usually neutral or feebly alkaline, but in consequence of fermentation rapidly becomes strongly acid in reaction, with the simultaneous formation of lactic, acetic, and traces of other volatile acids.

Qualitative Detection of Sugar.

49. Different methods serve for this purpose :—

(1) The sugar can be obtained in a crystalline form, providing it occurs in considerable quantity in the urine.

To this end, evaporate a portion of the urine to syrupy consistence, upon a water bath. The sugar separates from the solution upon standing, in yellow, warty masses, which by recrystallization can be further purified. Often there is found in urine a sugar which is perfectly uncrystallizable, and that remains in syrup form.

(2) Moore's test. Place a quantity of urine in a narrow, tolerably long test tube, add sodium or potassium hydrate, and heat the upper portion of the liquid. If sugar be present in rather large quantity, this part will assume a yellow, or brownish red color, while the lower layer will retain its original color.

(3) In doubtful cases the fermentation test is useful. A small portion of yeast is placed in a large test tube, and the latter then filled with the urine. The filled tube is inverted over a small quantity of water, or urine, and allowed to stand for some hours, the temperature ranging from 30–40°C. Any sugar present will break up into alcohol and carbon dioxide—

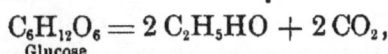

$$C_6H_{12}O_6 = 2\,C_2H_5HO + 2\,CO_2,$$
Glucose

and the resulting carbon dioxide will collect at the top of the tube. If the apparatus of Fresenius and Will be employed in performing the test, the sugar can be estimated quantitatively.

(4) Another test is to boil the urine under consideration for some time, with an ammoniacal silver nitrate solution. If there be any sugar present, the silver will deposit in metallic form, as a beautiful bright mirror upon the sides of the vessel. Formic and tartaric acids give a similar reaction.

(5) A solution of *indigo-carmine*, rendered alkaline by

75 parts dilute acetic acid (containing 30 per cent. acid).
120   "   water.

(9) Trommer's test. Mix the sample of urine (freed of
albumen) in a test tube, with a few drops of potassium or
sodium hydrate, warm *gently*, to expel any ammonia
present, filter if a large precipitate of earthy phosphates is
formed, and then, after cooling, carefully add drop by drop,
a dilute copper sulphate solution as long as the voluminous
precipitate first formed dissolves. *Heat the resulting clear
blue liquid gently*, and if sugar be present the solution will
soon become cloudy, and instead of the blue color, yellow
striped separations are noticed, which increase gradually
until finally the entire liquid assumes a yellow color. On
standing for a little time a yellow precipitate of hydrated
cuprous oxide or of red cuprous oxide separates. Boiling
should be avoided when heating the liquid. The heating
of the urine with the alkaline hydrate before the addition
of the copper solution should be very *gentle*, otherwise,
when only traces of sugar are present, it can be so *altered*
that it will not reduce the copper. If the alkaline copper
solution and the urine are heated to boiling, the copper
can be reduced by organic matters that are present and
sugar be entirely absent. The inexperienced, therefore,
should make repeated tests.

If the preceding mixture of urine and alkaline copper
solution *is not heated at all*, but left standing perfectly *cold*
for 12–24 hours, if sugar be present, cuprous oxide will
separate. (The other organic substances in urine only
reduce the copper solution on the application of heat.)

(10) Fehling's test. About 5 c.c. of Fehling's solution
are poured into a test tube and brought to boiling. This
should always be done before adding the suspected urine,

for the reason that by standing for some time Fehling's solution undergoes decomposition, which unfits it for making the sugar test. If upon boiling a precipitate should form, the solution should not be used; on the other hand, if no precipitate is formed, proof is shown that no change has taken place, and that the solution is reliable. The suspected urine is then added, drop by drop, and, if sugar be present, the blue color will change to green, and almost immediately to yellow, hydrated cuprous oxide or red cuprous oxide being formed. If only minute quantities of sugar be present, several cubic centimetres of the urine may be required to give the reaction.

Blitz brings out sharply and elegantly the well known reaction between a solution of sugar and Fehling's liquid by mixing with the latter a concentrated sodium chloride solution, heating to boiling, and carefully adding to this a sample of the urine under examination. The strong sodium chloride solution prevents a mixture of the two liquids, so that at their point of contact the red coloration appears with great distinctness.

Seegur (Centralblatt für die Med. Wissenschaften, 1875, p. 323,) has confirmed the assertion that a solution rather rich in sugar will reduce Fehling's solution in the cold. This property is absent when the sugar is present in but minute quantities. He found that an aqueous sugar solution, containing 0.1 per cent. sugar produced scarcely any reduction in the cold; that an aqueous sugar solution containing 0.05 per cent. sugar will not produce any reduction whatever in the cold. An artificially prepared sugar solution containing 0.1 per cent. sugar caused in the cold a very slight decolorization of the copper solution without any separation of cuprous oxide. A sugar solution

of the same strength (0.1 per cent.) after filtration through
animal charcoal, was found entirely without action in the
cold, while when warmed it caused the most beautiful
separation of cuprous oxide. Experiments with pure uric
acid solutions indicated that the same when containing as
little as 0.5 per cent. uric acid reduced Fehling's solution
very rapidly in the cold.

Maly (Sitz. d. k. Akad. der Wissenschaft. März Heft,
1871,) has found that 28 milligrams of a 1 per cent. creatinin solution dissolved the cuprous oxide furnished by 10
milligrams sugar (1 per cent. solution).

To detect sugar when contained in small quantities in
urine, and also to free the latter from creatinin, Bence
Jones employs the following modification of Brücke's
method: To 50 cubic centimetres of urine add 60 cubic
centimetres of lead acetate solution (strength 10 per cent.),
filter, and to the filtrate add lead basic acetate as long as a
precipitate forms, filter. again, and to this last filtrate
add ammonium hydrate. Collect the precipitate formed
by the ammonium hydrate on a filter, wash thoroughly
with water, remove with a horn spatula from filter paper
and suspend it in water; through this mixture pass a
stream of hydrogen sulphide. Filter off the precipitated
lead sulphide, boil the filtrate, to expel the hydrogen
sulphide remaining in solution, and after evaporating to a
bulk equal to the original volume of urine employed or less,
apply the tests for sugar.

Another method proposed by Carnelutti and Valente
(Gazz. Chim. x. 473-475) for the removal of creatinin is as
follows: 100 c.c. of urine, decolorized by passing through
animal charcoal, are evaporated to a syrup and mixed with
1 c.c. of a solution composed of 25 per cent. zinc chloride,

25 per cent. hydrochloric acid, 50 per cent. water. To the syrup, after the addition of the zinc chloride mixture, is added double the volume of alcohol, filtered, after standing several hours, the filter paper washed with alcohol, the alcoholic filtrate evaporated, and the residue diluted with water to the original volume of urine employed, and with this liquid the tests for sugar can be made. Fehling's quantitative method can be performed without any of the cuprous oxide going into solution. Loss of sugar does not take place in the performance of the above method.

Small quantities of carbolic acid do not, but larger ones do affect the reaction of sugar with bismuth subnitrate. Carbolic acid also interferes in the test with Fehling's solution. Readily oxidizable substances, such as the hypophosphites, hinder the coloration of the sugar by potassium hydrate on warming (Moore's test), but hasten, apparently, the reduction of bismuth and copper. Hyposulphites also hasten the reaction with the bismuth, but deport themselves differently with Fehling's solution. On boiling the latter with hyposulphites, the blue color remains unaltered, and is decolorized on the addition of the sugar solution without separation of cuprous oxide. After standing awhile there is deposited a black mass, consisting mostly of copper sulphide. Chloral, added to an alkaline solution of sugar and bismuth subnitrate, is rapidly decomposed, chloroform and formic acid are produced, and the reduction of the bismuth will be delayed until all the chloral is decomposed.

Quantitative Determination of Sugar in Urine.

50. (1) By fermentation. The carbonic acid apparatus of Fresénius and Will is employed here. It consists of two flasks connected by means of a glass tube bent twice at

right angles.   In one of the glass vessels we place about 30 c.c. urine, together with some well washed yeast and a small quantity of tartaric acid. The apparatus is properly arranged, then weighed, and afterwards placed where there is a temperature of 20° to 30° C.   In a short time fermentation sets in.   The generated carbon dioxide passes through the sulphuric acid in the second flask and escapes into the air. In three days the fermentation is complete.   The apparatus is then warmed gently and weighed when cool.   The loss in weight, due to the escape of carbon dioxide, multiplied by 2.045 will represent the amount of sugar present in the given volume of urine.

This method can be considerably modified, at least the apparatus can be dispensed with, by taking the specific gravity of a given volume of urine, adding a little yeast, allow it to ferment and again determine its specific gravity. Multiply the loss sustained by .23, or divide by 4.37, and the product will be the percentage amount of sugar present in the urine employed.

(2) With the standardized alkaline copper solution. (The so-called Fehling's solution.)

This is prepared as follows :—

It is found that one molecule of sugar exactly reduces the copper in five molecules of copper sulphate, therefore, in order to make a copper solution in which we have sufficient copper sulphate to be exactly equivalent to five grams of sugar, we use the following proportion :—

$$180 \quad : \quad 1247.5 \qquad :: 5 : x$$
$$C_6H_{12}O_6 : 5\,CuSO_4 + 5H_2O :: 5 \text{ grms. sugar} : x = 34.6525$$

grams crystallized copper sulphate, which are to be dissolved in 200 c.c. water.

173 grams chemically pure crystallized sodium potas-

sium tartrate (Rochelle salts) are dissolved in 480 c.c. sodium hydrate solution of 1.14 specific gravity. To this we now gradually add, with constant stirring, the copper sulphate solution, and the mixed clear liquid is diluted with distilled water to one litre. Of this solution

1000 c.c. = 5. grms. sugar.
10 c.c. = .050 grm. sugar.

10 c.c. of this copper solution will be reduced by 0.050 grm. grape sugar.

The above copper solution can only be preserved for a time without decomposition, by filling in small vessels of from one to two ounces capacity, which are then closed with tight fitting corks, sealed with wax or paraffin, and kept in a cool, dark cellar. Or the copper sulphate and double tartrate solutions can be kept in separate, well corked bottles and mixed in proper proportion just before being used for analysis. However, it is always best to boil a sample of the Fehling's solution before using, to make certain that no decomposition has taken place, so that the copper will be reduced even in the absence of sugar.

51. To make a sugar determination by this method a quantity of urine is diluted with nine or nineteen times its volume of water, and then placed in a burette. We now take 10 c.c. of the Fehling's solution, place it in a flask, or porcelain dish, dilute it with 40 c.c. water, and heat the mixture to boiling; then allow the diluted urine to run in from the burette until all the copper has been reduced to cuprous oxide. This point is recognized when, after standing some time, the cuprous oxide subsides, and the vessel held towards the light shows a colorless supernatant liquid. A filtered portion of this liquid acidified with acetic acid should not give a precipitate with ferrocyanide of potas-

sium, nor with hydrogen sulphide. Another filtered portion is boiled with a few drops more of Fehling's solution. If a precipitate be formed in either of the first two tests, the reduction is not complete, and more urine must be added; if the few drops of Fehling's solution added to the other portion be reduced, too much urine has been added, and the whole operation should be repeated.

In making the test, it is advisable to heat the copper solution to gentle boiling, over a spirit lamp, or Bunsen burner, and when the solution assumes a red color, remove the flask, or dish, to allow the cuprous oxide to subside. The nearer the point of complete reduction, the more rapidly will the precipitate subside. As this determination is rather difficult for the inexperienced, it should be repeated several times.

Albumen, as previously indicated, must be removed by coagulation and filtration. The calculation is as follows:—

Suppose we diluted 10 c.c. of urine with 190 c.c. of water, and of this diluted liquid 25 c.c. were required to reduce the 10 c.c. of Fehling's solution, then we would have—

$$200 \; : \; 10 \; :.: \; 25 \; : \; x$$
$$x = \frac{10 \times 25}{200} = \frac{250}{200} = 1.25 \; \text{c.c.}$$

and in these 1.25 c.c. urine are contained 50 milligrams of sugar. From this we calculate how much sugar was eliminated in twenty-four hours. If a diabetic patient voided about 5000 c.c. urine, then we would have this proportion:

1.25 c.c. : .050 milligram : : 5000 c.c. : x = 200,000 milligrams, or 200 grams of sugar.

(3) Knapp's Method yields results that agree perfectly with those obtained by the preceding method, and further,

possesses decided advantages in the easy preparation and preservation of the mercuric cyanide solution employed. 400 milligrams of the mercury salt require 100 milligrams of grape sugar for complete reduction; 10 grams dry and pure mercuric cyanide are dissolved in enough water to effect solution, 100 c.c. of sodium hydrate solution of 1.145 sp. grav. are added, and the whole diluted with water to one litre. In making an analysis, place 40 c.c. of the mercuric cyanide solution in a flask, and heat to boiling. Now run in the urine so diluted as to contain about one-half per cent. sugar; all the mercury is precipitated. In the quantity of the urine mixture required for the complete reduction, there must have been exactly 100 milligrams of sugar.

On adding the sugar solution to the boiling alkaline mercuric cyanide solution, the latter will become immediately turbid, but clears again towards the end of the operation and assumes a yellow color. To follow the course of the method, moisten a strip of Swedish filter paper, from time to time, with a drop of the mixture, and then with a glass rod bring a drop of ammonium sulphide close to the spot for about one-half minute. The whole spot at first becomes brown, but toward the end only its edge presents a clear brown ring, which may be noticed only by holding the transparent spot towards a bright light. Finally, the fresh, transparent spot is wholly unchanged by the ammonium sulphide, so that with some practice the $\frac{1}{10}$ c.c. of a one-half per cent. sugar solution can be easily titrated. For complete satisfaction, filter finally a few c.c. of the liquid, acidify with acetic acid and test with hydrogen sulphide for mercury.

(4) By polarization. The so-called observation tube is filled with clear, filtered urine, not containing any albumen,

taking care, also, to prevent the inclosure of any air bubbles, and then placed in Mitscherlich's or Ventzke-Soleil's
polarization apparatus. Notice accurately on the scale
and the verniers of the instrument, the rotatory power, and
from this calculate the quantity of grape sugar by means
of the formula—

$$p = \frac{a}{+56} \, 1,$$

in which p represents the quantity of sugar in grams for
1 c.c. of urine; a, the observed rotation; 1, the length of
the observation tube, and $+56$, the specific rotation. This
method requires frequent practice, in order to obtain accurate results.

Suppose we had, for example, found that the plane of
polarization had been turned 3.5 to the right, then the
equation would be—

$$56 : 100 : : 35 : x$$
$$\frac{100 \times 35}{56} = 6.25$$

therefore, a rotation of 3.5 degrees would indicate 6.25 per
cent. sugar.

INOSITE IN URINE.

52. Inosite has been found constantly in urine in Bright's
disease and albuminuria, in uræmia after the use of dras-
tics, in *diabetes mellitus*, in two cases of carcinoma, and
once in the urine of a convalescent from cholera. In one
instance of diabetes the inosite gradually displaced the
sugar originally present. Külz (Centrallblatt f. d. med.
Wissensch., 1876, p. 550.) has confirmed the assertion of
Strauss according to whom inosite is a constituent of normal urine, whenever there has been excessive drinking of

water. It may be detected as follows: Urine from which albumen has been completely removed is saturated with lead acetate solution, filtered, and the concentrated filtrate mixed with basic lead acetate as long as a precipitate appears. The latter contains the inosite combined with it.   The precipitate is collected on a filter paper and well washed with water, and then scraped off and suspended in water, and a stream of hydrogen sulphide passed through. The precipitated lead sulphide is filtered off. The filtrate from the lead sulphide may deposit some uric acid.   This can be filtered off, the filtrate concentrated quite considerably, and while boiling mixed with three to four times its volume of alcohol. Should this produce a heavy precipitate which tends to adhere to the sides of the vessel, then pour off the alcoholic solution; but if there is only a flaky turbidity, filter through a warmed funnel, and allow the solution to cool.   In about twenty-four hours the inosite will separate out from the filtrate in cauliflower-like grouped crystals.

Inosite is insoluble in alcohol and ether, readily soluble in water. The aqueous solution has a sweet taste. Yeast does not decompose inosite into alcohol, but decaying cheese will effect this. It is further recognized in its behavior toward nitric acid. On evaporating it with nitric acid to dryness, and moistening the residue with a little ammonium hydrate and calcium chloride, and again evaporating, a brilliant rose-red coloration results. A transparent gelatinous mass, which soon becomes starch-like in appearance, is produced on warming an inosite solution with basic lead acetate. The reaction with mercuric nitrate is also worthy of note.

LACTIC ACID AND LACTATES.

53. Lactic acid has been observed in urine in the acid fermentation, and results, very likely, from the decomposition of urinary extractive and coloring matters. It is also asserted that this acid has been found in urine when there was obstruction of the oxidation in the blood, therefore, in disturbances of respiration, digestion and nourishment in the urine of rachitic children and in leucæmia.

As lactic acid does not present any marked chemical properties, its zinc salt, which crystallizes readily in characteristic forms—mallet-shaped—is used to detect it. The urine intended for its preparation should be as fresh as possible. Inasmuch as its occurrence in urine is very variable, and it does not afford any definite diagnosis, we can omit the remaining properties.

FATS AND FATTY ACIDS.

54. Fat is very rarely found in urine. It has been noticed in the fatty degeneration of the kidneys (Bright's disease), in the fatty degeneration of the epithelial cells of the urinary organs and bladder, and in excessive chylous, or fatty blood serum (urina chylosa, cause unknown). From time to time, of the volatile acids, butyric has been found, and in combination in fermented diabetic urine, acetic and propylic acids. Owing to the small quantities of fat in urine, it is very difficult to detect by chemical means. The microscope affords us the best solution of the problem, as the fat globules appear here as flattened round plates of remarkable refracting power, and dark, tolerably irregular contours. When it is impossible to recognize the fat under the microscope, the urine under examination is

evaporated upon a water bath, the residue dried for some
time at 110° C., then extracted repeatedly with ether.
When the ether has evaporated, only fat remains, and its
presence can now be confirmed under the microscope, and
by its deportment toward heat (acrolein) and paper (grease
spots).

BILIARY COLORING MATTERS, BILIARY SALTS AND TAURIN.

55. Although biliary coloring matters are likely to occur
in healthy persons during the hot portions of the year, such
instances are rare.   Both biliary coloring matters and
salts, and now and then taurin (decomposition of tauro-
cholic acid), are found in icterus.   Urine charged with
biliary pigments is easily recognized by its decided tinge
of color, being at one period red-brown, and then grass
green.   Such urine foams strongly when shaken, and colors
filter-paper yellow or green.

For the detection of either of the above we employ nitric
acid.   Place in a test tube some concentrated and slightly
yellow-colored nitric acid, and then carefully add, by
means of a pipette, some of the urine under examination,
taking care that the two liquids do not intimately mix.   In
the presence of biliary pigments there will be produced at
the junction of the two liquids a beautiful play of colors,
at first a *beautiful green* ring, which gradually rises higher,
exhibiting slowly at its lower surface a *blue, violet, red* and
finally *yellow* ring (green is characteristic for bile pigments).

Urine containing bile, when treated with hydrogen per-
oxide, ferric chloride, and an acetic or phosphoric acid
solution of lead superoxide, shows a beautiful green color.

Masset (Journ. de Pharm., et de Chim., [4], 30, 49),

employs the following modification of Gmelin's test for the detection of biliary pigment in urine. 2 cubic centimetres of urine are acidified with 2–3 drops concentrated sulphuric acid, and then a small crystal of sodium nitrite introduced into the liquid. In the presence of bile pigments magnificent grass-green streaks appear, which, on shaking, color the entire liquid dark green. This color does not disappear on boiling and remains many days unaltered. Even traces of biliary coloring matter produce a distinct pale green coloration.

Traces of bilirubin are detected by shaking the urine with chloroform, which becomes yellow in color If nitric acid (containing nitrous acid) be poured on the chloroform the play of colors mentioned before as produced with nitric acid will be noticed.

To detect the acids of the bile (of which cholic acid is the starting point), separate the sodium salts from the urine and treat the concentrated aqueous solution with from 2 to 3 drops of sugar solution (1 to 4) then add a little pure concentrated sulphuric acid. The liquid is at first turbid, then it becomes clear and almost at the same moment yellow, then pale cherry red, *dark carmine red*, and finally beautiful purple violet.

LEUCIN, TYROSIN AND CYSTIN.

56. Leucin and tyrosin have been found in acute yellow atrophy of the liver, in typhus, variola, and in the urine of an epileptic after injury to the spinal cord. Urine containing cystin has frequently been observed. The relation of cystin to any definite changes in disease has not yet been determined. When leucin and tyrosin are abundant in urine, they can easily be detected. Tyrosin is either al-

ready found crystallized out, or it separates simultaneously with leucin on evaporating the urine to a small bulk and allowing it to cool, when the well known characteristic forms are microscopically recognized (leucin in brown, oily-like layers, tyrosin in sheaf-like needles). If the quantity of these substances is not so abundant that they appear upon the evaporation of the urine, the method of Frerichs should be pursued. A rather large quantity of urine, usually rich in biliary pigments and albumen, is precipitated with basic acetate of lead, filtered, the excess of lead in the filtrate removed by passing a stream of hydrogen sulphide through it, and the filtered and clear solution reduced to a small volume on a water bath. If tyrosin is present, in twenty-four hours it will be found nicely crystallized out; while the leucin, being much more soluble, separates later.

OCCURRENCE OF FIBRIN IN URINE.

57. Fibrin very rarely occurs in urine, and is of little definite diagnostic importance. When it is found present we are justified in the conclusion that there has been a fibrinous transudation from the blood into the kidneys or urinary passages. The presence of fibrin is characterized by the formation of fibrinous coagula some hours after the urine has been voided. These coagula deposit as a sediment or convert all the urine into a gelatinous mass. The microscope will show the regular *fibrin cylinder* as rolled-up with sharp contour and yellow or brown yellow color. (Plate III, Fig. 2.)

BLOOD PIGMENTS IN URINE.

58. Blood pigments have been detected in urine in cer-

tain diseases which accompany dyscrasia and blood degeneration, in scurvy, in putrid typhus fevers, in pernicious alternating fevers, and after the inhalation of hydrogen arsenide.

In these instances the urine is bloody, colored from *redbrown* to ink black. Yet a microscopic examination will not reveal the elemental forms of the blood.

Upon boiling such urine alone, or after the careful addition of some drops of acetic acid, a brown-red coagulum is formed, which, with alcohol containing sulphuric acid, yields hæmatin.

BLOOD IN URINE.

59. In troubles induced by the presence of calculi in the bladder or kidneys, causing a mechanical lesion of certain vessels, or in violent desquamative nephritis, finally, in severe cystitis in which the texture of the bladder suffers, blood can occur as such in the urine. It can, in addition, occur as a result of the effusion of the blood into the urinary canal, that, by the coagulation of the blood of the urethra the passage for the urine will be obstructed so that the voiding of the urine will be impaired, or that these coagula will induce the formation of permanent concretions in the urinary channel. If blood be present in the urine, fibrin and albumen, as integral parts of the blood, will also be found, and, therefore, it will be very necessary to proceed carefully if we wish to ascertain whether all the albumen occurring in the urine originated from the effused blood or from other sources.

Almen (Neues Jahrbuch für Pharmacie, 40 p. 232,) recommends the following for the detection of blood in urine. Mix in a test tube some drops of tincture of guaiacum with an equal volume of oil of turpentine, and shake until an emulsion forms, then carefully add the urine under examination, so that it falls to the bottom of the tube. On agitating the emulsion with the urine, the guaiacum resin is rapidly precipitated as a white, afterward dirty yellow or green precipitate. If there be blood in the urine, and even if only in traces, the resin is colored a more or less intense blue, often almost indigo blue in color. In normal, albuminous, or urine containing pus, this blue coloration does not occur, but only appears in the presence of blood.

HYDROGEN SULPHIDE IN URINE.

60. Hydrogen sulphide is very rarely observed in urine. It has been noticed in the so-called reabsorbed urine, and its occurrence attributed to the fusion of the exuded proteids. According to Beetz, under certain conditions ammonium sulphide can reach the blood from the skin, and there produce phenomena of poisoning similar to those observed in the inhalation of sewer gas. In this case the urine yields tests for ammonia and hydrogen sulphide. In violent cystitis from the decay of albuminous urine in the bladder, hydrogen sulphide will be formed, and it is then a very unfavorable prognosis. The detection of it is easy—the odor distinguishing it. A slip of paper moistened with lead acetate is immediately blackened when immersed or held over urine containing hydrogen sulphide. In this test the urine should be slightly warmed.

OXYMANDELIC ACID.

61. O. Schultzen and L. Riess discovered oxymandelic acid as an abnormal constituent in urine, together with leucin, tyrosin, and sarco-lactic acid. Its formula is $C_6 H_8 O_4$.

F

The urine in which this occurs contains also biliary pigments, biliary acids, albumen in traces, and that peptone-like substance which is noticed in urine in considerable quantities after phosphorus poisoning. The urea is either perfectly absent, or present in diminished quantity.

To obtain the acid, free the urine, by evaporation, from tyrosin and leucin, precipitate the mother liquor with alcohol, evaporate the alcoholic solution, and the syrupy residue, after the addition of dilute sulphuric acid, is exhausted completely with ether. The united ethereal extracts leave upon evaporation a brown, liquid residue, from which long, thin, colorless needles separate, which are then dissolved in water and the solution filtered. In the feebly yellow-colored filtrate lead acetate produces only a slight flocculent precipitate, which decolorizes the liquid. With basic lead acetate, the filtrate gives a voluminous flocculent precipitate, which condenses to a heavy, granular, crystalline powder. This compound is suspended in water and decomposed by hydrogen sulphide. On evaporation the filtrate deposits colorless, silky, very flexible needles—constituting the new acid.

INDICAN.

62. Indican has recently been found to be an indoxyl-sulpho acid. Jaffè estimates it quantitatively by means of bleaching lime. 1000 to 1500 c.c. of urine are made alkaline with calcium hydrate and the phosphates then removed by means of calcium chloride. Filter after twelve hours, evaporate the filtrate to a thick syrup. The syrupy residue is warmed some minutes with about 500 c.c. alcohol, then brought into a beaker and allowed to stand twelve or twenty-four hours. Filter and distill off the alcohol. The

residue is dissolved in a large quantity of water and precipitated with a very dilute solution of ferric chloride. The filtrate from the iron precipitate is mixed with ammonium hydrate, boiled, and after filtration evaporated to 200-250 c.c. With this solution the determination is made. First determine the amount of chloride of lime necessary to separate the indigo. To this end measure out 20–40 c.c. of the liquid, and dilute this gradually with definite amounts of water, until 10 c.c. of the mixture treated with an equal volume of hydrochloric acid show a perceptibly blue coloration on the addition of a drop of a saturated bleaching lime solution. Multiplied experiments have shown that the number of volumes of dilution which can be added to an indican solution until the appearance of the limit of the reaction, is about double the number of drops of the bleaching lime solution, which will show the maximum indigo yield for 10 c.c. of indican. When the right proportion has been determined, mix 200–300 c.c. of urine with bleaching lime and hydrochloric acid, allow to stand at least twelve hours, collect upon a filter that has been extracted with hydrochloric acid and washed, dried, and weighed. Dissolve out the hippuric and benzoic acids with water, wash the residual indigo with dilute ammonium hydrate, and finally with water, dry, precipitate and filter at 105 to 110° C., and weigh.

V.

URINARY DEPOSITS (SEDIMENTS).

63. Urinary deposits are solid, undissolved substances in the urine which at first are mostly suspended in the latter, but after shorter or longer periods form a precipitate. Some are produced after, others before, the urine has

been voided. In the latter case they may form in the tracts of the urine (urinary organs, bladder, etc.), and under favorable circumstances produce *calculi.*

Many urinary sediments whose constituents were at first dissolved, separate or form in consequence of the peculiar alterations of the urine. These have already been described in preceding paragraphs, under the name of *acid* and *alkaline urine* fermentation.

64. The microscope is an indispensable aid in the examination of these deposits. Without it we would, in many instances, not be capable of arriving at a correct, conclusive decision. By its assistance we distinguish the various sedimentary forms, classing them as amorphous, crystallized and organized bodies. These are, however, not alike for every reaction of the urine. So far as the organized bodies are concerned it is only partially correct, while the occurrence of the crystalline and amorphous bodies is dependent in part upon the reaction.

65. Depending on the reaction in urine we find various substances in the deposits:—

A. In acid urine, are present:—

  (a) *amorphous bodies* : urates, phosphates and fats.
  (b) *crystalline bodies:* calcium oxalate, uric acid, calcium phosphate, cystin, tyrosin, hippuric acid.
  (c) *organized bodies:* mucous coagula, mucous corpuscles, pus, blood corpuscles, urinary casts, epithelial cells, fermentation and thread fungi, vibrionæ, spermatozoids, cancerous tissues, sarcina ventriculi Goodsir.

B. In alkaline urine are found :—

  (a) *amorphous bodies:* calcium carbonate, calcium phosphate.

(b) *crystalline bodies:* magnesium ammonium phosphate (triple phosphate), ammonium urate, crystallized calcium phosphate.

(c) *organized bodies:* in addition to the above, infusoriæ and confervæ (fermentation and thread-like fungi are increased).

Therefore, before beginning a microscopic examination, observe whether the urine has been newly voided, the reaction whether alkaline or acid. This done, the sediment is allowed to subside, the supernatant liquid decanted, and by means of a pipette, a drop of the sediment is placed on a glass slide, covered with a glass circle, and then brought under the objective of a microscope. Move the slide about, until all points have passed the field of vision. Having examined one sample, take a second, and be it noticed here that this specimen be taken from different layers of the sediment, inasmuch as some substances deposit more rapidly than others, and many, like calcium oxalate, only after the expiration of several hours. If filtration had been necessary for the separation of the deposit, be careful, in cleaning the filter paper, not to bring any fibres of the paper under the microscope and consider them solid constituents of the sediment.

I. The urine reacts acid.

A. *The entire sediment is amorphous,* presenting partially irregular masses, partially moss-like intertwined series, consisting of extremely fine grains.

Carefully warm the drop on the object glass.

(a) Perfect solution follows = urates. As a confirmatory test, add, after cooling, a drop of hydrochloric acid and allow to stand from one-quarter to one-half hour. If rhombic tablets of uric acid form in ·

this time, proof is sufficient. Usually this sediment consists of *acid sodium urate*, and is distinguished by a more or less red color.

(b) The sediment does not dissolve on the application of heat, but dissolves in acetic acid without effervescence = calcium phosphate. Chemically, the calcium is proven by ammonium oxalate ; the phosphoric acid by ammonium hydrate, and magnesium sulphate forming ammonium magnesium phosphate, or by means of ammonium molybdate.

(c) Beneath the sediment are found drops which refract light strongly = fat.

B. *The sediment, or deposit, contains well formed crystals :—*

(a) Calcium oxalate. Minute, shining, perfectly transparent, quadratic octahedra in the form of envelopes, which refract light strongly (Plate II, Fig. 2). As these crystals are light in weight, they deposit very slowly, and can readily be overlooked by the inexperienced. The urine should be allowed 12–24 hours to deposit and then be carefully decanted.

(b) Uric acid, following its principal form, crystallizes in rhombic tablets with rounded, blunt corners, which are known as Wetzstein's form (Plate I, Fig. 5). The crystals may be very small and some very complicated, building themselves upon accidental impurities, *e. g.*, threads, and forming series of hairs and long cylinders.

Again, the crystals are greatly developed, and united to a nucleus, when they appear upon the edge (fan-like), or upon the plane (shingle-like). Uric acid has also been found in cask shapes and long spears, combined with rosettes. Owing to coloring matters precipitated at the same time, the uric acid is either pale yellow, or brown-red to dark-brown (Plate I, Figs. 4–5).

For chemical confirmation, see page, 32.

(c) The crystallized calcium phosphate, viewed under the microscope, presents either individual keel-shaped crystals, or several are arranged in regular order, so that they are with their sides towards each other, and their ends converging to one point. In addition, perfect circular rosettes are found, and sometimes the crystals not only arrange themselves in circles, but build parts of spheres. The urine usually reacts feebly acid.

(d) Cystin forms regular, six-sided tables, soluble in ammonium hydrate and hydrochloric acid. They carbonize and burn on being heated. Boiled with a sodium hydrate solution of lead oxide, lead sulphide is produced. The chemical test for cystin consists in this last experiment, and that on being heated on platinum foil it does not fuse, but burns with a greenish blue flame and the diffusion of an odor very similar to prussic acid.

Cystic urine is generally pale. The assertion has been made that, frequently, several individuals of the same family will suffer from cystinuria.

(e) Tyrosin forms delicate short needles, which cross each other so frequently that they present a sheaf-like appearance, of which every two sheaves super-impose in the form of a cross. Chemically the tyrosin crystals are tested according to the method of Piria, or Hoffmann. According to the first, a minute quantity of the sediment is placed on a watch

glass and moistened with two to three drops of sulphuric acid. In about half an hour, add a little
water, neutralize with sodium carbonate, as long as
it effervesces, then filter. If the sediment was tyrosin, the solution, on the addition of neutral ferric
chloride, will show a violet color. Hoffmann's
method is simpler. Pour water over a portion of
the sediment, boil, and add to the boiling liquid a
few drops of mercuric nitrate, when a red precipitate will form, while the supernatant liquid is colored
rose to purple red.

Urine containing tyrosin frequently contains biliary pigments.

(f) Hippuric acid, as a sediment, rarely occurs. It
crystallizes in needles and rhombic prisms, soluble
in water. (See Plate I, Fig. 3.)

C. *The sediment contains organized bodies:—*

(a) Mucous coagula, forming wound-up strips, consisting of serrated, very minutely arranged points and
grains, frequently accompanied by sodium urate.

(b) Mucous Corpuscles. Very small, contracted and
granulated corpuscles, generally combined at the
edges to large shield-like groups.

Large quantities of mucus can form in urine,
without affecting the transparency of the latter.
Only on protracted standing, when there commences
a deposition of urates, or when the urine contains
more epithelium than usual mixed with it, does the
mucus become visible, as a cloud. If the turbidity
disappears on the application of heat, the urates were
the cause. Small crystals of calcium oxalate and
uric acid, as well as individual mucous corpuscles,

or epithelium of the bladder, which bodies had been suspended in the mucus, have also been found.

(c) Blood corpuscles form circular, slightly bi-concave disks, generally with a yellow appearance, again reddish with a faint touch of green. They greatly expand by acetic acid, dissolving in this more or less slowly. (See Plate III, Fig. 3.)

Particular attention should be directed to swollen, spherical and also distorted zigzag forms (readily produced by a concentrated sodium sulphate solution).

In the presence of blood the urine contains albumen. To detect blood pigments in urine, precipitate the earthy phosphates of the urine in a test tube with potassium hydrate, warming gently. In the precipitation the phosphates carry down the pigments, appearing not white, as in normal urine, but blood red. When but a small amount of blood pigment is present in the urine, the earthy phosphates show dichroism.

(d) Pus. Round, pale, granulated cells of varying magnitude, usually as large again as blood corpuscles, increasing markedly when touched with acetic acid, and losing their granulated surface and giving rise to residues of varied forms and groupings. It is impossible, either chemically or microscopically, to distinguish these corpuscles from mucous corpuscles, but in the presence of pus the urine always contains albumen. (See Plate III, Fig. 4.)

By Donné's test the pus in urine can be detected without the assistance of a microscope. To do this, pour off the urine from the sediment, add a small piece of solid potassium hydrate to the latter, and stir some minutes with a glass rod. If the sediment consists of pus, it will be deprived of its white color, becoming greenish and glassy, at first thready, finally more compact, until eventually it results in a coherent body, *i. e.*, it has assumed the appearance peculiar to pus in strong ammoniacal urine. Only in case the quantity of pus was small, it cannot be expected to result in a compact lump, but the sediment may be made to disappear, and a thready, gluey liquid results.

(e) Urinary casts are tube-like cylinders, often accompanied by blood and pus corpuscles, holding in their substance or walls epithelial cells and mucous corpuscles.

(a) The epithelial casts of the Bellini tubes, whose round cells are distinctly visible as a delicate molecular mass.

(β) Granulated kidney casts are of granular, cloudy appearance.

(γ) Hyaline kidney casts are solid, of paler, more transparent appearance. Often distinguished from the surrounding liquid with only the greatest difficulty. (See Plate III, Fig. 1.)

(f) Epithelial cells in their different forms, dependent on their origin. (See Plate II, Fig. 6.)

(1) Squamous epithelium. Round, longitudinal or polygonal cells from the major and minor labiæ and the vagina, from the female urethra, the bladder, the kidneys. (See Plate III, Fig. 6.)

(2) Cylindrical and spheroidal epithelium from the lower layer of the mucous membrane of the bladder.

PLATE II.

FIG. 1.

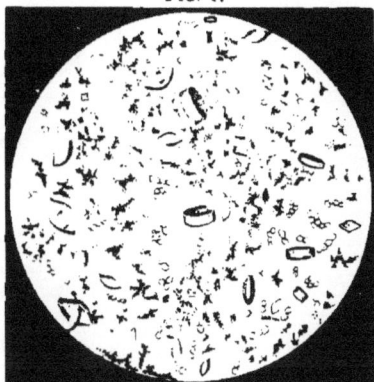

Uric Acid, Sodium Urate and fermentation fungi.

FIG. 2.

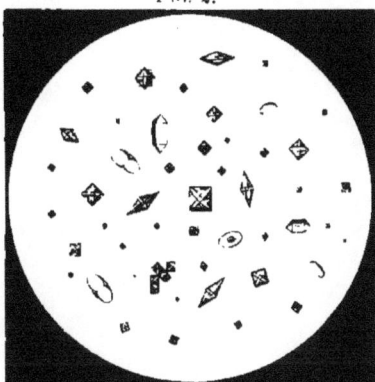

Calcium Oxalate.

FIG. 3.

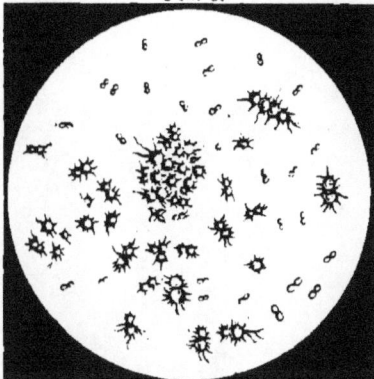

Ammonium Urate.

FIG. 4.

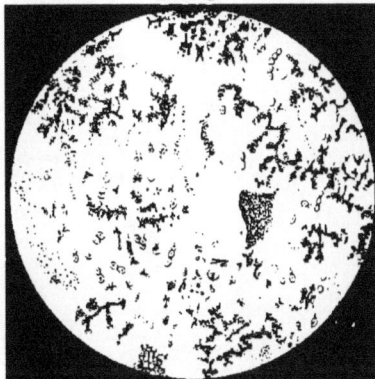

Sodium Urate, Phosphates and Coagulated
Mucin.

FIG. 5.

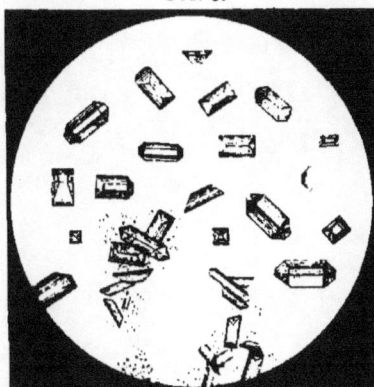

Triple Phosphates.

FIG. 6.

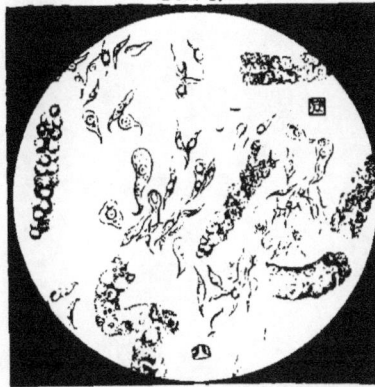

Epithelial casts and Epithelial cells.

(3) Glistening columnar epithelium from the uterus. (Addition of iodine solution makes all these formations more distinct under the microscope).

(g) Fermentation and Thread-like Fungi. In the first stage of the acid urine fermentation they accompany the sediments of sodium urate, free uric acid and calcium oxalate, but are found most frequently in diabetic urine, and such as has passed into fermentation.

    (1) The fermentation fungi form small nucleated cells, which increase by formation of sprouts, and thus form simple or intertwined series.

    (2) Thread-like fungi often produce so thick a tissue that they obscure the field of vision.

(h) Vibrionæ are short, delicate rods, moving actively hither and thither (under high power observed in feebly acid and alkaline urine).

(i) Spermatozoids. Microscopic, somewhat elongated, pear-shaped bodies, with a more or less long, hair-like tail, which may or may not be in constant vibration. They are found—

    (1) After coition. When a portion of the seminal fluid had remained in the urethra and was discharged in the urine later.

    (2) In spermatorrhœa. Besides the independent disease of this name, involuntary emissions of seminal fluid have been noticed in serious cases of typhus.

(j) Cancerous masses:

    (1) Distinct cancer cells.

    (2) Small pieces of cancerous tissues.

    The first are often unusually large, most

frequently having, apparently, a cilium with very large, often multiplied, nuclei. Care must be taken not to confound the ciliated cells originating in the pelvis of the kidney with the cancerous cells. The superstructure of the villous cancer consists of dendritic vegetation, upon which sometimes the epithelial growth rests. Such masses are voided spontaneously with the urine from the bladder. Again, it is only after examination, as, for instance, in the introduction of the catheter, that they are loosened and appear subsequently in the urine. (See Plate III, Fig. 5.)

(k) Sarcina ventriculi Goodsir. Very rare. The characteristic form is not readily confounded with anything else.

II. The urine reacts alkaline.

A. *The sediment contains amorphous bodies.*

(a) In alkaline urine these consist only of calcium phosphate.

B. *The sediment contains crystals.*

(a) The ammonium magnesium phosphate occurs usually in combinations of the rhombic, vertical prisms, in equal coffin-lid-like crystals, which acetic acid dissolves easily (distinction from calcium oxalate), and on warming with sodium hydrate, ammonia gas is liberated. (See Plate II, Fig. 5.)

(b) Ammonium urate consists of brown colored spheres, which are developed singly, or every two are combined to double spheres, presenting entire conglomerations with reniform surface. The latter is smooth or set with small points like a thorn apple, or the projecting points are long, evenly divided, and then mostly bent, which gives rise to a great multiplicity of intermingled forms. (See Plate II, Fig. 3.)

Ammonium urate, like other urates, gives the murexide test.

C. *The sediment contains organized bodies:*

Besides blood, mucus and pus corpuscles, fermentation and thread-like fungi, infusoriæ and confervæ are found.

Relations of Sediments to the Diagnosis of Disease.

66. (1) Uric acid and urates occur not only in pathological urine, in acute, febrile diseases, but also in normal urine. In newly voided urine sediments of free uric acid never occur, except in renal calculus, while on the other hand every urine in the course of acid fermentation deposits uric acid crystals. The deposits of urates, especially potassium and sodium urates, are very frequent, and represent the fever sediments (sedimenta lateritia), long known to physicians. They are sometimes deceptively similar to mucus, pus and blood, and are only recognized by their microscopic character.

(2) Deposits of calcium oxalate occur in both healthy and diseased individuals. Oxaluria, which is the name applied when they occur abundantly, is of great diagnostic importance, although it occurs in some other diseases, as dyspepsia, spermatorrhœa and diseases of the spinal cord. In oxaluria the urine is dark in color. (See Plate II, Fig 2.)

(3) Hippuric acid deposits are found frequently after the eating of fruit, the ingestion of benzoic and cinna-

mic acids, and in various diseases. It has very little diagnostic importance.

(4) The rarely observed sediments of cystin are of little diagnostic value. Generally present in renal calculus.

(5) Tyrosin sediments have been observed in acute liver diseases.

(6) Sediments of ammonium magnesium phosphate (triple phosphate) are found constantly, when the urine, because of the conversion of urea into carbon dioxide and ammonia, becomes alkaline.

(7) Calcium phosphate occurs under the same conditions.

(8) Mucus corpuscles (mucin) are constantly present in traces in normal urine, also in febrile conditions of the most varied type, as pneumonia, pleuritis, typhus, respiratory and intestinal catarrh, meningitis, etc.

(9) Tube casts are observed in many diseases, particularly Bright's disease of the kidneys. They constitute the principal basis in the diagnosis and prognosis of certain diseases of the renal parenchyma.

(10) Spermatozoids exist in urine after pollution or coitus ; also not unfrequently in the urine of typhoid patients. They point to an unusual and decidedly excessive irritation of the genital organs.

(11) Fungi and infusoriæ in freshly voided urine indicate that it has decomposed in the bladder, which is tolerably often the case in catarrh of the bladder.

(12) Pus in urine always indicates suppuration in the uropoetic system, or points to an abscess related to the latter. The question of importance is, is the pus the product of a superficial affection of the mucous membrane (catarrhal inflammation), or of a graver affec-

tion of this membrane, intimately connected with material alterations? To answer the question observe the continuance of the suppuration, and the properties of the pus.

(13) Cancer and tubercular masses show the presence of cancerous or tubercular depositions which have softened in almost any part of the uropoetic system: example, cancer of the bladder, and rarely, cancer of the kidneys.

VI.

PRACTICAL HINTS TO A COURSE FOR THE QUALITATIVE AND QUANTITATIVE EXAMINATION OF URINE.

67. As a rule, it is scarcely necessary to examine for all the normal and abnormal substances in urine. Proof of the presence of one or several of the mentioned constituents is sufficient for diagnosis, and it is only where the physician desires an accurate knowledge of all the nourishment relations of an individual, that it can be of value to him to extend the analysis to all substances found in the urine. In such instances, a single analysis is insufficient, only a series of repeated analyses being satisfactory.

The substances that are to be looked for dictate the course of analysis.

In examining urine, we always regard it as pathological, consequently, we search for abnormal constituents. An exception would be urine containing a sediment. Here examine both the liquid and the deposit, and class the substances found as first (a) in the sediment, (b) in solution.

Of course, some of the normal constituents should be searched for, such as salts, etc., and in the report of the examination, catalogue the various ingredients found under the headings normal constituents, and abnormal constituents. This is advisable for the practitioner, because he does not always retain in memory the various constituents of urine, so that from an arbitrary arrangement of the normal and abnormal constituents, he is able to present a clear picture. This is more readily accomplished when the detected substances are arranged under the mentioned headings. This is advantageous, too, where an accurate examination may be required.

The physician having determined the substance to look for, whose chemical detection is principally concerned, the same is sought under the respective headings, and tested as therein directed.

If, on the contrary, a general examination is desirable, pursue the plan recommended by Neubauer.

68. *Qualitative course.*

I. Determine the reaction with litmus.
   The urine may be :—
   (a) acid and clear.
   (b) acid and sedimentary.
   (c) neutral or alkaline. In the latter case, a deposit is usually present. The filtered urine, free from sediment, is further tested. (Section 65.)

II. Albumen. Biliary pigments and blood. Heat a small quantity of the urine (if it does not give an acid reaction), with addition of a drop or two of acetic acid, to boiling. The formation of a coagulum, not removed by nitric acid, indicates albumen. If the coagulum is—
   (a) *white:* it consists of pure albumen. (See page 52.)
   (b) *greenish:* there is good reason to suspect biliary

PLATE III.

FIG. 1.

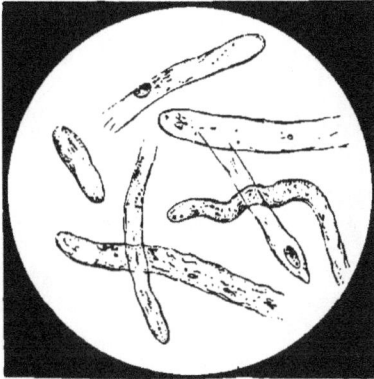

Hyaline Cast.

FIG. 2.

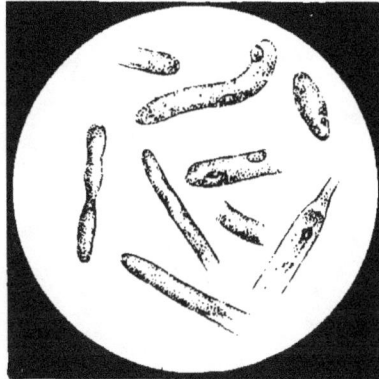

Fine granulated casts.

FIG. 3.

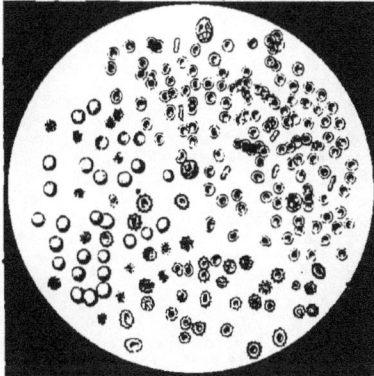

Blood corpuscles.

FIG. 4.

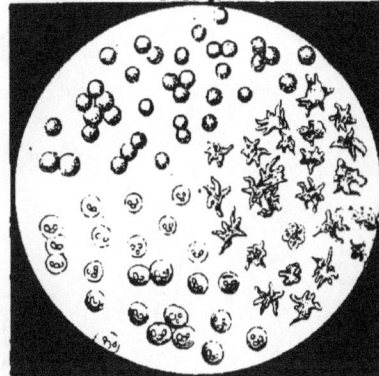

Pus corpuscles

FIG. 5.

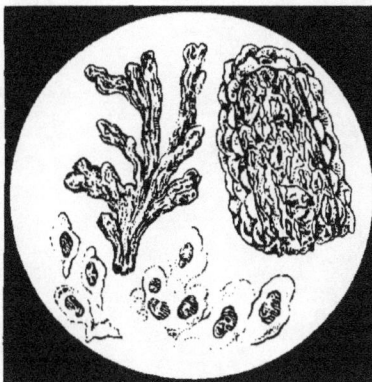

Organized growth found in urinary sediment from an individual having cancer in the bladder.

FIG. 6.

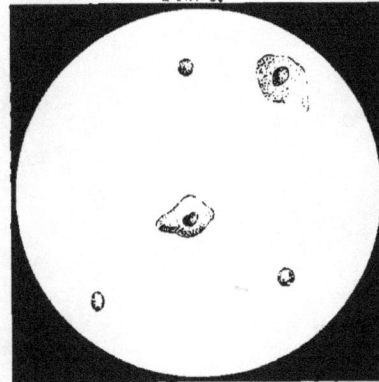

Sediment from normal urine, showing several mucus corpuscles (young cells) and squamous epithelia.

pigments, especially if the urine be highly colored. (See page 71.)

(c) *brownish-red:* blood may be present. (See page 72.)

III. Urea. Creatinin. Uric acid. Hippuric acid. Lactic acid. Earthy phosphates, etc. About 400 to 500 c.c. of clear urine, free from sediment and albumen coagulum, are evaporated upon a water bath to thick syrupy consistence, and then divided into two parts ($\frac{1}{3}$ and $\frac{2}{3}$).

(1) $\frac{1}{3}$ of the residue is exhausted with strong alcohol; allow the undissolved portion to subside, filter, wash . the residue again with strong alcohol and test the solution according to *a* and *b*, the residue according to 3.

(a) *Urea.* A small portion of the alcoholic liquid is evaporated almost to dryness on a water bath, the residue is dissolved in a little water, and a few drops of pure nitric acid free from nitrous acid (as this decomposes the urea into carbon dioxide, water and nitrogen), or oxalic acid, added, to strongly acid reaction. Upon cooling, urea nitrate or oxalate separates in white shining scales, or hexagonal tablets, the oxalate sometimes in four-sided prisms. (See Plate I, Fig. 2.)

(b) *Creatinin.* $C_4H_7N_3O$. Mix the greater portion of the alcoholic solution with a few drops of calcium hydrate, and then add calcium chloride as long as a precipitate is produced. Filter, reduce the filtrate on a water bath to 10 or 12 c.c., then pour this into a beaker, and after cooling, add $\frac{1}{2}$ c.c. of a pure alcoholic solution of zinc chloride. The precipitate collected after some hours' standing is examined

G

microscopically. (It usually forms delicate needles concentrically grouped, giving rise either to perfect rosettes or tufts.)

(2) Hippuric acid. $C_9H_9NO_3$. The two-thirds portion of the residue in III, feebly acidulated with hydrochloric acid, is triturated with heavy spar powder (barium sulphate), and exhausted with alcohol. The alcoholic extract is saturated with sodium hydrate, the alcohol distilled off, and the syrupy liquid, after the addition of oxalic acid (to combine with the urea) evaporated to dryness on a water bath. Powder the residue and treat with ether, distilling off the latter and treating the warm residue to remove the excess of oxalic acid with calcium hydrate. Filter and reduce the filtrate to a small volume and acidify with a little hydrochloric acid. After a short time hippuric acid will crystallize out and can be examined chemically and microscopically. If the residue is gluey lactic acid is indicated. (See page 68.) If upon pouring some of the ethereal solution on water the well-known fat phenomena show upon the surface, fat is present.

Hippuric acid crystallizes from a hot solution in delicate needles, from a cold saturated solution in milk-white, perfectly transparent four-sided prisms and columns, having two to four planes upon their extremities; the principal form is a vertical prism (see Plate I, Fig. 3). (Distinction from benzoic acid, which crystallizes in tablets overlying each other.) On fusing, hippuric acid becomes first an oily fluid, and on cooling solidifies to a white crystalline mass, and this on being further heated to almost glowing, leaves a porous coke in addition to sublimed benzoic acid and ammonium benzoate, and liberates an odor strongly similar to that of hydrocyanic acid. Strong nitric acid dropped into boiling hippuric acid, and evaporated to dryness leaves a residue, which, if heated in a small glass tube, sets free, like benzoic acid, an intense odor of oil of bitter almonds, from the formation of nitro-benzene.

(3) The residue obtained in 1 is placed in a dish and dilute (1 part acid and 6 parts water) hydrochloric acid poured over it. The portion remaining undissolved is collected on a small filter.

(a) The earthy phosphates and other salts are found in the hydrochloric acid filtrate, and are precipitated by the addition of an excess of ammonium hydrate.

(b) The residue on the filter contains mucin and uric acid. After washing with water, pierce the filter and with a stream of water from a wash bottle wash the residue into a small test tube, add 2 to 3 drops of sodium hydrate, warm and filter.

(a) The undissolved residue is mucin.

(β) The filtrate will contain the uric acid, which yields crystals on being mixed with hydrochloric acid.

IV. Urine coloring matters. (See page 15.)

V. Glucose. (Test, page 55.)

VI. Hydrogen sulphide. (See page 73.) The urine smells of it, and colors paper saturated with lead acetate brown or black.

VII. Inorganic substances.

Evaporate 40–50 c.c. of urine to dryness on a water

bath. Mix the residue with one to two grams of spongy platinum and ignite gently until all the carbon is burned off, when a greenish white mass remains. Reserve a small portion to test for iodine (see IX), the rest boil with water and obtain—

A, a solution, and
B, a residue.

A. The solution is divided into four parts and examined for :—

(1) *Sulphuric acid.* Acidify one part with hydrochloric acid and add barium chloride; a white pulverulent precipitate, insoluble in acids.

(2) *Chlorine.* Acidify a second portion with nitric acid and add silver nitrate ; a white curdy precipitate that blackens on exposure to light.

(3) *Phosphoric acid.* Mix a third portion with sodium acetate, acetic acid, and add a few drops of ferric chloride ; yellowish-white gelatinous precipitate. Another test is with ammonium molybdate, the liquid in presence of phosphoric acid is colored yellow and a yellow precipitate is produced.

(4) *Sodium.* The remainder of the solution is evaporated to dryness and a small portion of the residue heated on a platinum wire in the inner flame of a blowpipe; yellow coloration imparted to flame.

(5) *Potassium.* A portion of the residue in 4 is dissolved in a little water and platinum tetrachloride added ; a yellow crystalline precipitate.

B. The residue is extracted with hot hydrochloric acid, filtered, washed, and the filtrate examined for :—

(1) *Iron.* Heat a portion with a drop of nitric acid, and add sulphocyanide of potassium; deep red coloration.

(2) *Calcium.* Add an excess of sodium acetate to a second portion, and test with ammonium oxalate; white precipitate.

(3) *Magnesium.* Precipitate all the calcium as in 2; filter and add ammonium hydrate to filtrate; there is formed a white precipitate of ammonium magnesium phosphate.

VIII. Ammonium salts. 50–100 c.c. of urine are mixed in a flask with sodium hydrate, and above it, by aid of the cork, a strip of moistened turmeric paper is hung. If ammonia is present, the paper rapidly becomes brown, or if a glass rod moistened with hydrochloric acid is held over the mouth of the flask, the well-known ammonium-chloride clouds are produced.

IX. Iodine. Use the reserved portion of VII, put it in a porcelain crucible, moisten it with some drops of fuming nitric acid, and place a little starch paste on the under side of the lid of the crucible, which is then covered over the latter. In presence of iodine the starch is colored violet. The original urine containing iodine can also be immediately distilled with sulphuric acid. The latter method is, however, more complicated.

H. Struve's colorimetric method for the estimation of iodine depends on a color scale, prepared by taking a potassium iodide solution of known strength, carbon bisulphide and a few drops of fuming nitric acid, and with this the color of the iodine solution from the urine is compared.

Iodine can be estimated quantitatively by the following method : 10–20 c.c. of palladious chloride, depending on the quantity of iodine in the urine, learned by previous qualitative tests, are heated in a corked flask, upon a water bath, and the urine containing iodine acidified with hydrochloric acid, and reduced by evaporation to a definite volume (10–20 c.c.) added, until all the palladium is precipitated as palladious iodide. Agitation of the mixture hastens the separation. Small portions of the urine filtered off from time to time, and added to the urine under examination, show when the reaction is complete.

X. Regarding the examination for the less important constituents of the urine, such as phenol, for the detection of which 20–30 kilos. of urine must be employed ; further, benzoic and acetic acids, which only occur in decomposed urine; we can omit them, referring for details to larger works upon this subject. The same may be remarked of—

XI. Butyric Acid. which is rarely found, and requires several kilos. of urine to detect it.

XII. Inosite. (See page 66.)

XIII. Allantoin.

XIV. Xanthin.

XV. Leucin and Tyrosin. (See page 70.)

The quantitative determination of the various constituents has already received treatment.

VII.

URINARY CONCRETIONS.

(*Urinary Gravel and Calculi.*)

69. Concretions of the urine are deposits from the urine within the tracts (kidneys, ureter, bladder and urethra). Sometimes they are as small as grains of sand, and are consequently voided with the urine without much inconvenience. In such cases they are very abundant, and, as a rule, crystalline (urinary sand—gravel). Frequently they are larger, varying from the size of a pea to that of a small apple, and then cannot be voided (the true calculi). A sharp line of distinction between the two cannot be drawn; generally, they are distinguished by difference in their form.

70. The calculi consist mostly of a homogeneous mass, or of several concentric layers, frequently of chemically distinct substances, which have arranged themselves around a nucleus (very often a dried particle of mucin), and here gradually increased.

71. We recognize and distinguish them readily under the microscope, especially where sand or particles have accidentally fallen into vessels and been mistaken for calculi by hypochondriacal patients. The grains, consisting mostly of silicates, are distinguished, by their appearance and physical deportment, from calculi, and rarely is a chemical examination necessary.

Chemical Constituents of Calculi.

72. They are essentially identical with those already mentioned under urinary sediments, and for their closer

examination the student is referred to the preceding pages. Calculi may consist of :—

(1) Uric acid and urates.

(2) Xanthin.

(3) Cystin.

(4) Calcium oxalate.

(5) Calcium carbonate.

(6) Calcium phosphate.

(7) Ammonium magnesium phosphate.

(8) Proteid substances.

(9) Urosteatite mixed with considerable quantities of silicic acid, aluminium oxide, etc.

73. In making a chemical examination pursue the following course :—

After careful microscopic examinations (in calculi the different layers), which are important, because many calculi consist of but one of the above constitutents, pulverize the object to be examined, wash off with a little cold distilled water, dry and ignite a sample upon platinum foil, over a Bunsen burner or spirit lamp.

I. Either no or a very slight residue remains.

II. The calculus appears to be incombustible or leaves a large residue after ignition.

74. When there is no residue, or at most but a slight one, the following substances may be present :—

Uric acid,
Ammonium urate, } burn without flame.

Xanthin,
Cystin,
Urosteatite,
Proteid substances (as Fibrin, etc.), } burn with flame.

The chemical tests for these bodies are:—

(1) *Uric acid.* Test by treating the powder with nitric acid and ammonium hydrate (murexide). Calculi of uric acid are relatively very frequent, and can attain a large size. Generally they are colored (yellow, reddish and reddish-brown), rarely white, and possess usually a smooth surface and considerable hardness.

(2) *Ammonium urate.* A portion of the sample treated with potassium or sodium hydrate liberates ammonia gas, recognized by white clouds formed about a glass rod, moistened with hydrochloric acid.

Uric acid and ammonium urate are distinguished by the fact that uric acid is only slightly soluble in water, while ammonium urate dissolves much more readily, and in larger proportion. Calculi of ammonium urate are rare and generally of small size, of a clear (white or clay-yellow) color, and rather earthy appearance.

When the murexide reaction is not obtained the combustible concretion may consist of:—

(3) *Xanthin.* Soluble in nitric acid without liberation of gas. In evaporating the solution there remains a residue of intense lemon yellow color, not reddened by ammonium hydrate, but soluble in sodium or potassium hydrate, with a deep reddish-yellow color.

Guanin, has not yet been detected in urinary calculi. It yields a similar reaction to xanthin, therefore, care is necessary here.

Calculi of xanthin are very rare, and thus far have been found in few instances. They have a clear brown (white to cinnamon-brown) color, are tolerably hard, with a waxy lustre, acquired by rubbing, and consist of concentric, easily soluble amorphous layers.

(4) *Cystin* dissolves in ammonium hydrate, and crystallizes by spontaneous evaporation from this solution in very characteristic crystals, forming regular, six-sided tables, which occasionally are attached to a large six-sided rosette.

On dissolving a calculus containing cystin in potassium hydrate, and boiling after the addition of a small quantity of lead acetate, there is formed a black precipitate of lead sulphide, which imparts to the mixture an inky tint.

Cystin calculi are also very rare, of pale yellow color and smooth surface, with crystalline fracture and waxy, or greasy lustre. They are moderately soft, easily shaved, and the powder formed is much like that of soap.

(5) *Proteid substances* do not exhibit the slightest trace of crystallization, diffuse, on burning, the odor of burning horn, insoluble in water, ether or alcohol, soluble in potassium hydrate, from which solution they are precipitated by acids. In acetic acid they expand and swell up, and are soluble in boiling nitric acid.

Calculi from proteid substances (formed from fibrin and blood coagula) are very infrequent.

(6) *Urosteatite* fuses when heated, without effervescence, swells up and liberates a very strong odor, recalling that of a mixture of shellac and benzene. It dissolves in potassium, or sodium hydrate, with saponification. Very soluble in ether. The residual urosteatite, after the evaporation of the ethereal solution, becomes violet on further warming.

Calculi of this kind, like the preceding, are extremely rare. In the fresh condition they are soft, elastic, and resemble caoutchouc; on drying, they diminish in size, become brittle, light-brown to black, are moderately hard, and become softer on warming.

75. If the calculus did not burn, or left a large residue after ignition, it may consist of sodium, calcium, or magnesium urates, oxalate and carbonate of calcium, ammonium magnesium phosphate, and calcium phosphate.

76. As we have already described the chemical tests of these substances, we will confine ourselves to only the most important points in what follows:—

(1) *Sodium, calcium, and magnesium urates* do not readily occur as the sole constituents of calculi. Yet they sometimes are present in greater or less quantity in the calculus. To ascertain whether uric acid is united with such a base, boil the powder with distilled water, and filter while hot. The urates, more soluble in warm water than uric acid, pass into the filtrate. This is evaporated, ignited, and the bases in the residue tested for by the various previously described methods. For the insoluble uric acid, see preceding pages.

(2) *Calcium oxalate* blackens on ignition, by its conversion into calcium carbonate. Continued strong ignition leaves calcium oxide. Calculi of calcium oxalate are rather frequent, especially in children. They are either small, pale colored and smooth, *Hemp-seed calculi*, or they are larger, of rough exterior, bunchy, warty, colored dark brown on their surface and sometimes even black, *Mulberry calculi*. Owing to their rough surfaces, they irritate the urinary passages and induce serious disorders (bleeding, inflammation).

(3) *Calcium carbonate.* Easily recognized by its effervescence with acids; blackens also on ignition, resulting from organic substances present.

(4) *Ammonium magnesium phosphate* and (basic) calcium phosphate occur generally intermixed with each

other; do not burn on ignition, but fuse to a white, enamel-like mass; hence called fusible calculi. After strong ignition they never react alkaline, differing in this respect from calculi of calcium oxalate and carbonate. In hydrochloric acid they dissolve without effervescence, and are re-precipitated from such an acid solution by ammonium hydrate.

(5) In very rare cases calculi of neutral calcium phosphate occur. In their chemical and physical properties they resemble the earthy phosphates, but differ from these in not containing any magnesium.

COMPOSITION OF A SAMPLE OF URINE.

ANALYSIS BY MILLER.

Water,	956.37	
Urea, . .	14.23 ⎫	
Uric acid, .	0.37 ⎪	
Extractive matter.	15.05 ⎬	29.81 organic matter.
Mucus. . .	0.16 ⎭	
Sodium chloride, .	7.22 ⎫	
Phosphoric acid, .	2.12 ⎪	
Sulphuric acid, .	1.70 ⎪	
Calcium oxide, .	0.21 ⎬	13.82 inorganic matter.
Magnesium oxide, .	0.12 ⎪	
Potassium oxide, .	1.92 ⎪	
Sodium oxide, . .	0.53 ⎭	
	1000.00	43.63 total solids.

TABLE FOR THE TENSION OF AQUEOUS VAPOR FOR TEMPERATURES FROM —2° TO 30°, CELSIUS, (BUNSEN).

°C.	Tension in Millimet'rs	°C.	Tension in Millimet'rs	°C.	Tension in Millimet'rs	°C.	Tension in Millimet'rs
—2.0	3.955	6.2	7.095	14.4	12.220	22.6	20.389
—1.8	4.016	6.4	7.193	14.6	12.378	22.8	20.639
—1.6	4.078	6.6	7.292	14.8	12.538	23.0	20.888
—1.4	4.140	6.8	7.392	15.0	12.699	23.2	21.144
—1.2	4.203	7.0	7.492	15.2	12.864	23.4	21.400
—1.0	4.267	7.2	7.595	15.4	13.029	23.6	21.659
—0.8	4.331	7.4	7.699	15.6	13.197	23.8	21.921
—0.6	4.397	7.6	7.840	15.8	13.366	24.0	22.184
—0.4	4.463	7.8	7.910	16.0	13.536	24.2	22.453
—0.2	4.531	8.0	8.017	16.2	13.710	24.4	22.723
0.0	4.600	8.2	8.126	16.4	13.885	24.6	22.996
+0.2	4.667	8.4	8.236	16.6	14.062	24.8	23.273
0.4	4.733	8.6	8.347	16.8	14.241	25.0	23.550
0.6	4.801	8.8	8.461	17.0	14.421	25.2	23.834
0.8	4.871	9.0	8.574	17.2	14.605	25.4	24.119
1.0	4.940	9.2	8.690	17.4	14.790	25.6	24.406
1.2	5.011	9.4	8.807	17.6	14.977	25.8	24.697
1.4	5.082	9.6	8.925	17.8	15.167	26.0	24.988
1.6	5.155	9.8	9.045	18.0	15.357	26.2	25.288
1.8	5.228	10.0	9.165	18.2	15.552	26.4	25.588
2.0	5.302	10.2	9.288	18.4	15.747	26.6	25.891
2.2	5.378	10.4	9.412	18.6	15.945	26.8	26.198
2.4	5.454	10.6	9.537	18.8	16.145	27.0	26.505
2.6	5.530	10.8	9.665	19.0	16.346	27.2	26.820
2.8	5.608	11.0	9.792	19.2	16.552	27.4	27.136
3.0	5.687	11.2	9.923	19.4	16.758	27.6	27.455
3.2	5.767	11.4	10.054	19.6	16.967	27.8	27.778
3.4	5.848	11.6	10.187	19.8	17.179	28.0	28.101
3.6	5.930	11.8	10.322	20.0	17.391	28.2	28.433
3.8	6.014	12.0	10.457	20.2	17.608	28.4	28.765
4.0	6.097	12.2	10.596	20.4	17.826	28.6	29.101
4.2	6.183	12.4	10.734	20.6	18.047	28.8	29.441
4.4	6.270	12.6	10.875	20.8	18.271	29.0	29.782
4.6	6.350	12.8	11.019	21.0	18.495	29.2	30.131
4.8	6.445	13.0	11.162	21.2	18.724	29.4	30.479
5.0	6.534	13.2	11.309	21.4	18.954	29.6	30.833
5.2	6.625	13.4	11.456	21.6	19.187	29.8	31.190
5.4	6.717	13.6	11.605	21.8	19.423	30.0	31.548
5.6	6.810	13.8	11.757	22.0	19.659		
5.8	6.904	14.0	11.908	22.2	19.901		
6.0	6.998	14.2	12.064	22.4	20.143		

ADDENDA.

Professor Wormley's paper (page 30) can be found in the *American Journal of Medical Sciences*, July, 1881, page 128.

INDEX.

103

N. B.—For uran. acetate, p. 43, write—$(C_2H_3O_2)_2\, UrO_2 + 3H_2O$